PRESENTING AMERICA'S WORLD

Re-materialising Cultural Geography

Dr Mark Boyle, Department of Geography, University of Strathclyde, UK and
Professor Donald Mitchell, Maxwell School, Syracuse University, USA

Nearly 25 years has elapsed since Peter Jackson's seminal call to integrate cultural geography back into the heart of social geography. During this time, a wealth of research has been published which has improved our understanding of how culture both plays a part in and – in turn – is shaped by social relations based on class, gender, race, ethnicity, nationality, disability, age, sexuality and so on. In spite of the achievements of this mountain of scholarship, the task of grounding culture in its proper social contexts remains in its infancy. This series therefore seeks to promote the continued significance of exploring the dialectical relations which exist between culture, social relations and space and place. Its overall aim is to make a contribution to the consolidation, development and promotion of the ongoing project of re-materialising cultural geography.

The series will publish outstanding original research monographs and edited collections which make strong and clear contributions to the furtherance of the re-materialisation agenda. Work which foregrounds the role of culture in shaping relations of domination and resistance will be particularly welcomed. Both theoretically reflexive contributions charting the progress and prospects of the agenda, and theoretically informed case studies will be sought:

1) The re-materialising agenda – progress and prospects: including the location of the agenda within the broader development of human geography; reflexive accounts of the main achievements to date; outstanding research agendas yet to be explored; methodological innovations and new approaches to field work; and responses to the challenges posed by non-representational theory and theories of performativity.

2) Theoretically informed case studies within the tradition: including work on new links between culture, capital and social exclusion; changing concepts of masculinity and femininity; nationalism, cosmopolitanism, colonial and post-colonial identities, diaspora and hybridity; the rescaling of territorial identities, the new regionalism and localism, and the rise of supranational political bodies; sexuality and space; disability and the production of and navigation around the built environment.

Presenting America's World

Strategies of Innocence in *National Geographic Magazine*, 1888-1945

TAMAR Y. ROTHENBERG

Bronx Community College of the City University of New York, USA

First published 2007 by Ashgate Publishing

Published 2016 by Routledge
2 Park Square, Milton Park, Abingdon, Oxfordshire OX14 4RN
711 Third Avenue, New York, NY 10017, USA

First issued in paperback 2016

Routledge is an imprint of the Taylor & Francis Group, an informa business

British Library Cataloguing in Publication Data
Rothenberg, Tamar Y.
Presenting America's world : strategies of innocence in
National Geographic magazine, 1888-1945. -
(Re-materialising cultural geography)
1. National Geographic Society (U.S.) - History 2. National
Geographic magazine - History 3. Geography - United States
- Periodicals - History
I. Title
910.9'73

Library of Congress Cataloging-in-Publication Data
Rothenberg, Tamar Y.
Presenting America's world : strategies of innocence in National Geographic magazine, 1888-1945 / by Tamar Y. Rothenberg.
p. cm. -- (Re-materialising cultural geography)
Includes bibliographical references and index.
ISBN 978-0-7546-4510-8
1. National Geographic--History. 2. American periodicals--History. I. Title.

G1.N275R68 2007
910--dc22

2007003825

ISBN 13: 978-1-138-27659-8 (pbk)
ISBN 13: 978-0-7546-4510-8 (hbk)

Contents

List of Plates *vii*

Acknowledgements *ix*

Introduction: Pictures of Innocence 1

1 *National Geographic* in the New World Order 25

2 Picturing the World, Imagining the Nation 41

3 Picturing Human Geography: Orders of Science and Art 69

4 Maynard Owen Williams: Contradictions of a 'Seeing-Man' 99

5 Harriet Chalmers Adams: Intricacies of Class, Gender and Gusto 131

Afterword 165

Bibliography *171*

Index *185*

List of Plates

3.1 "Types in La Paz," ran the caption from February 1909 88
3.2 "A daughter of a dying [Marquesan] race – beautiful, luxuriant hair, fine eyes, perfect teeth, a slender, graceful form, a skin of velvet texture and unblemished figure 91
4.1 Seeing-man: Maynard Owen Williams in the rigging of the Arctic-bound *Bowdoin*, June 20, 1925 101
4.2 "My feeling is for the beautiful and very often the beautiful as suggested by an arch of lightly leaved bamboo bent so it frames a cloud," Williams wrote to his wife Daisy 102
4.3 Maynard Owen Williams believed that "by restricting its articles to what is kindly," *National Geographic* "made friends the world round" 105
4.4 "A Cobbler of Liangchow," subject of a debate over photographic representations 107
4.5 "Island, Plain and Mountain Furnished These Costumes for Athenian Maids" 111
4.6 "The Face with the Smile Wins" 129
5.1 "You could not obtain a more popular, entertaining, or instructive lecturer than Mrs. Harriet Chalmers Adams" 136
5.2 "It seemed incredible that the small, charming woman, gowned in deep red velvet with a long train, could have visited such strange places" 145
5.3 "A Woman Street-Car Conductor of Valparaiso" 149
5.4 "Ex-head-hunting country, Philippines" 152
5.5 "A Bedouin Family Rides to Town on its Two-Seater, Desert Model, Roadster" 159

Acknowledgements

Perhaps it goes back to the dolls. My grandparents, Harry and Emma Mann, spent many post-retirement years traveling, bringing me back dolls in what presumably was a form of native dress. The dolls – plastic, porcelain, cloth, paper – had personalities as disparate as their forms and costumes, but I was fascinated by the existence of all the different parallel worlds the dolls represented. So while this project is a critique of another popular form of representing people of different places and cultures, it is done with sympathy for human curiosity regarding difference. And I thank my grandparents for their part in piquing and sustaining my interest in the world and how we comprehend it.

There are many people to thank for making this book happen: encouragers, assistors, guiders, readers, editors. The project itself began at Rutgers University. Julie Tuason sparked my initial interest in examining early *National Geographic*, and her work on the magazine's coverage of the Philippines influenced my own inquiry. Neil Smith was greatly supportive in my launch in this arena, pushed me to grapple with theory and hone in on my thesis, and tugged me along through various renderings. Anne Godlewska helped sharpen my first excursion into this project, Briavel Holcomb and Philip Pauly offered consistently helpful critical readings, Peter Wacker, Bob Lake and Cindi Katz provided needed encouragement, and Rick Shroeder challenged me to critically examine the contemporary *National Geographic*. Phil Pauly graciously lent me not only his expertise, but his notes and files from his own work on the early years of the National Geographic Society. Briavel Holcomb lured me to geography in the first place, then put up with me (and put me up) for the next several years; I could not have done it without her.

Jillana Enteen, Alex Weheliye, Melina Patterson and Caroline Goeser gave me much needed editorial assistance, and Rupal Oza, Laura Liu, Lisa Lynch and Helen Hurwitz also combed through various bits and pieces. Additional support and inspiration came from many at Rutgers and beyond, especially Allan Frei, Sarah Thompson, Alyssa Katz, Ruthie Gilmore, John Kasbarian, Mike Craghan, James DeFilippis, James DeFilippis, Ted Killian, Craig Gilmore, Susan Blickstein, Shira Birnbaum, Sadat Obol, Cheryl Gowar, Don Mitchell, Maria Espinosa, Hong-Ling Wee, Allen Douglas and Roger Balm. Susan Schulten, who shared my interest in American popular geography, also shared her work and observations with me.

A fellowship at the Center for Critical Analysis of Contemporary Culture helped me focus on my writing, as did a fellowship from the Society of Woman Geographers. I thank the SWG for their support and for granting me permission to publish photographs from their collection of and by Harriet Chalmers Adams, the Society's first president.

Outsiders' access to the National Geographic Society Records Library is severely limited, but it was possible to request a file if I knew very specific information, such as a letter from person X to person Y on a particular day, or the names and dates of a particular assignment. Thanks to the considerate access I *was* granted, courtesy of Mary Ann MacMillan, I was able to see some letters and pamphlets held in the National Geographic Society Records Library. Ashley Morton and Barbara Moffet took care of all my National Geographic permissions.

I knew of some exact dates of letters because *National Geographic* staffer Maynard Owen Williams often noted his correspondences in a five-year Line-A-Day diary. That diary, along with other travel diaries, assorted manuscripts, photographs and correspondence are held in the Maynard Owen Williams Collection at Kalamazoo College in Kalamazoo, Michigan. Archivists Carol Smith and Elizabeth Smith gave me friendly and helpful assistance in my first visits there and Elizabeth Smith again helped me years later, in the midst of library renovations and archive dislocations.

I also thank the librarians and archivists of the Manuscript Division at the Library of Congress for their help, and Dana Martin, Lisa Lynch, Lyndsey Layton and Dan Mendelson for D.C. hosting and assistance.

I could not have written the Harriet Chalmers Adams chapter without Kate Davis, who pointed me in the direction of the Harriet Chalmers Adams Collection at the Stockton Public Library in Stockton, California, and who generously shared the early travel diaries she had discovered belonging to Adams's grand-niece and borrowed for her own research. I thank Kate not just for the materials but for the discussions and for her own influential interpretation of Harriet Chalmers Adams.

Thank you for encouraging me to turn the dissertation into a book: Bob Lake, Rick Shroeder, Neil Smith, Briavel Holcomb, Don Mitchell and thanks to Erik Dussere, Stephanie Hartman and Erik's mom for the early positive feedback. Thanks to Don Mitchell again for propelling me toward Ashgate. He and Mark Boyle offered suggestions for improving the manuscript that I hope I have done justice, though I suspect not. At Ashgate, thanks to Val Rose for her kind persistence and patience, and Neil Jordan and Carolyn Court, and in the important final stretch, Sarah Horsley.

My history department colleagues at Bronx Community College gave me considerable support and feedback. I benefited from our roundtable chapter treatment feel lucky to have landed in such a congenial place. Special thanks to Howard Wach for the discussions and to Sarah Danielsson for her trenchant reading of several chapters. And enormous thanks to Jackie Gutwirth, editor extraordinaire, whose sharp editorial judgments have surely made this a better book.

Alyssa Katz, Katie Gentile and Merryl Reichbach inspired me with their own book-writing efforts, and Helen Hurwitz offered ongoing support and encouragement in the final stages of the book.

Finally, I want to thank my family – Judy Rothenberg, Peter Rothenberg, the late Mark Rothenberg, and Richard Hazelton – for their love, support and forbearance. Rich nudged me on, explored Kalamazoo on his own while I pawed delicately over archives, formatted the photos, and proofread the entire manuscript. His eyes only glazed over once, and briefly.

For my parents

Introduction

Pictures of Innocence

Little should be needed to convince anyone living in the United States of the significance of *National Geographic* in American culture. Initiated in 1888 as a sporadic scientific journal and redirected as a non-technical monthly a decade later, by 1920 the *National Geographic Magazine* had a circulation of more than 750,000,[1] a number that for the most part only continued to rise. The *Geographic*'s success, declared an editor of the magazine in 1915, "proves how strong the love of this kind of geography is in every breast."[2] At the time of its centennial in 1998, the National Geographic Society had more than ten million members, each one receiving the magazine through the original member-subscriber arrangement. One did not subscribe to the *National Geographic*, but became a member of the National Geographic Society and received the magazine as a benefit of membership.[3]

By 1945, *National Geographic* was so ensconced in the American popular consciousness that it was used as a pivotal marker in the beloved American film *It's a Wonderful Life*. In the beginning of Frank Capra's 1946 film, the viewers, along with the character of Clarence the wingless angel, are permitted to see a few instructive episodes in the formative years of the protagonist, George Bailey. We first witness 12-year-old George Bailey saving his brother's life in 1919, a heroic deed that causes George to lose his hearing in one ear. We next see him months later working behind the counter of a drug store. Mary, the little girl who grows up to become his wife, is sitting on a stool at the counter and after some deliberation orders a chocolate ice cream. George suggests adding coconuts, but Mary tells him that she doesn't like them.

"You don't like coconuts!?" exclaims George. "Say brainless, don't you know where coconuts come from?" He puts down the ice-cream scoop and cup he is holding and reaches into his back pocket, whipping out a copy of the *National Geographic Magazine*. Placing it on the counter between himself and Mary, he says, "Look it says here – from Tahiti, the Fiji Islands, the Coral Sea –"

---

1 The Report of the Director and Editor of the National Geographic Society for 1919, Box 160, Grosvenor Family Papers, Manuscript Division, Library of Congress Washington D.C. [hereafter Grosvenor Papers]. The magazine's full title during the time period I examine was *The National Geographic Magazine*; it was shortened in the 1950s to *National Geographic*.

2 John Oliver La Gorce, *The Story of the Geographic* (Washington, D.C.: James Wm. Bryan Press, 1915). Unpaginated.

3 These days, one subscribes to the magazine; membership is a small-print accompanying benefit. See www.nationalgeographic.com.

Mary interrupts: "A new magazine!" She reaches for the *National Geographic*. "I never saw it before," she says, turning the magazine around and flipping through it.

"Of course you never," scoffs George as he grabs it back from her. "Only us explorers can get it. I've been nominated for membership in the National Geographic Society!"

He holds the magazine up, pointing proudly to the title of the magazine, then he bends over and busies himself with the ice cream, enabling Mary to lean over and whisper, "Is this the ear you can't hear on? George Bailey, I'll love you till the day I die."

George lifts himself back up, ice cream scoop in hand, and continues: "I'm going out exploring some day, you watch. And I'm gonna have a couple of harems, and maybe three or four wives. Wait and see." And he turns back to work, whistling.

I introduced this little scene because of its remarkable representation of *National Geographic*'s imprint on the American imagination.[4] Where coconuts come from is all of a piece with exploration and exotic male heterosexual fantasies, spouted by one of *National Geographic*'s archetypal readers, the All-American (that is, white, Christian, middle-class, small-town) boy. Geography is scripted as both knowledge and adventure, and as masculine prerogative. Young George scorns Mary's ignorance – "Say, brainless" – but ensures that she remain ignorant by grabbing the magazine back from her. The film uses *National Geographic* to signify George Bailey's lifelong dreams of unfettered travel, dreams that George cannot fulfill as he goes on to become a pillar of society as his small town's banker.[5]

It was not until the wake of the United States' 1898 plunge into overseas imperialism that *National Geographic* began its renowned embrace of "the world and all that is in it,"[6] basking in the curiosities uncovered and the economic potential of European and American imperialist endeavor. Banking on Americans' new interest in the world, the National Geographic Society became nationally expansive itself, shedding its form as a congenial association of like-minded and largely local scientists and emerging as a conduit of exploration, adventure, exoticism and natural history for the average American citizen.

How did *National Geographic* contribute to the way Americans thought about the world, and about themselves as Americans in relation to the rest of the world?

---

4 *National Geographic* influenced American cinema early on. Visiting Lasky Studios in Hollywood in 1921, Geographic librarian Kathleen Hargrave Frantz "found that the National Geographic Magazine was extensively used in obtaining data for decorations and costumes in moving-picture productions." Harry W. Frantz, from a conversation with Kathleen Hargrave Frantz recorded 7 May 1939, Box 267, Alexander Graham Bell Papers, Manuscript Division, Library of Congress. Washington, D.C. [hereafter Bell Papers].

5 Bankers were a favorite category of member-readers, at least in the magazine's appeals to advertisers. A full page ad supplement in early 1914, for example, trumpeted "18,000 Presidents and Vice-Presidents of Banks" among its readers.

6 C.D.B. Bryan, *The National Geographic Society: 100 Years of Adventure and Discovery* (New York: Harry N. Abrams, 1987), 43.

What did Americans see when they looked at the magazine? Who were some of the people who captured, in photos and text, parts of the world for that American reader? How did what these writers and photographers see cohere into the consistent vision of *National Geographic*? These are some of the questions I explore, examining the ways in which the magazine framed the world for its millions of readers during *National Geographic*'s formative first half century. I interrogate the ways in which the magazine became America's source of wholesome exotica and erotica, examine how it came to represent a particular form of gendered knowledge and activity, and consider its participation in the cultural work of U.S. global hegemony.

"The creation of consent is not a new art"

The concept of hegemony, as developed by Antonio Gramsci and interpreted by such writers as Stuart Hall and Raymond Williams, provides the framework for my examination of *National Geographic*.[7] Following Gramsci, Williams describes the concept of hegemony as one "which at once includes and goes beyond two powerful earlier concepts: that of 'culture' as a 'whole social process,' in which men define and shape their whole lives; and that of 'ideology,' in any of its Marxist senses, in which a system of meanings and values is the expression or projection of a particular class interest."[8] Hegemony implies fluidity; the dominant groups, an alliance of power elites, must continually work to maintain their dominance. They do so largely through the persuasion of subordinate groups. In this way, the system that advantages those already in power is rendered "common sense" and reasonable, if not desirable, to most everyone else. "Mastery is not simply imposed or dominative in character," notes Hall, but "results from winning a substantial degree of popular consent."[9]

Already in 1922, Walter Lippmann noted, "The creation of consent is not a new art." Much as Gramsci would do, Lippmann recognized the relationship between a democratic-styled state and what he termed the "manufacture of consent." Within his lifetime, he wrote, "persuasion has become a self-conscious art and a regular organ of popular government. None of us begins to understand the consequences, but it is no daring prophecy to say that the knowledge of how to create consent will alter every political calculation and modify every political premise."[10]

The mechanics of the manufacturing of consent came home to me during the 1991 Persian Gulf War. I participated in the enormous anti-war demonstration in

7 Antonio Gramsci, *Selections from the Prison Notebooks* (London: Lawrence & Wishart, 1971); Raymond Williams, "Selections from *Marxism and Literature*," in *Culture/Power/History*, ed. Nicholas B. Dirks, Geoff Eley and Sherry Ortner (Princeton: Princeton University Press, 1994): 585-608; Stuart Hall, "Gramsci's Relevance for the Study of Race and Ethnicity," in *Stuart Hall: Critical Dialogues in Cultural Studies*, ed. David Morley and Kuan-Hsing Chen (London: Routledge, 1996): 411-440.

8 Williams, "Selections," 595.

9 Hall, "Gramsci's Relevance," 424.

10 Walter Lippmann, *Public Opinion* (New York: Free Press, 1965), 158.

Washington, D.C. The U.S. government denied journalists access to the battle zone, but it was a *New York Times* editor, not the government, who buried an article about the demonstration on page 17 and featured a photograph of the ten anti-anti-war demonstrators rather than a photo of the thousands of people who braved the January cold to protest the U.S. invasion.[11] The *Times*' marginalization of the march proved to be but one small part of the larger construction of apparent national consensus in support of or acquiescence to the war in the mass media. The manufacturing of consent continued through constrained news reporting and nightly displays of target bombing – continually, in the case of CNN – on network and cable television.

While television was undoubtedly the most dominant and influential form of mass media in the late twentieth century, in the late nineteenth century, that role went to popular magazines. Part of the expression of consent is its very popularity. Huge numbers of people watch a television show or movie, or buy a magazine or record because these forms of cultural information and entertainment give them what they want. Magazine readers may absorb a specific editorial vision, but the effectiveness of that vision would be diminished if it did not find favor with its consuming readers. For *National Geographic*, the relationship between magazine and reader was even more intimate because readers were not simply subscribers but members whose dues helped pay for expeditions and so could take a degree of ownership in the subsequent reports in the magazine of adventures and scientific knowledge.

Popular magazines in the late nineteenth century such as *Munsey's*, *McClure's*, *Cosmopolitan* and the *Ladies' Home Journal* played an important role in the development of an American national culture, a process that continued well into the twentieth century. For Richard Ohmann, the most important aspect of this development was that an expanding professional-managerial class, which included both readers and producers of popular national magazines, learned to see themselves as consumers in a corporate capitalist system.[12] Matthew Schneirov also sees these magazines and their readers as participating in an adjustment to a new social order, though he emphasizes the ways in which the magazines intentionally addressed their readers as national citizens.[13]

Both studies make the point that the national magazines of the late nineteenth and early twentieth century contributed to a national imaginary, the "imagined community" of the United States, with an articulated national identity.[14] Whether it was *Ladies' Home Journal*'s 600,000 readers taking in a Pear's Soap advertisement in 1899 or *Cosmopolitan*'s 300,000 readers taking in the sights of the 1901 Pan

---

11 Peter Applebome, "Day of Protests Is the Biggest Yet," *The New York Times*, 27 January 1991, 17.

12 Richard Ohmann, *Selling Culture: Magazines, Markets and Class at the Turn of the Century* (London and New York: Verso, 1996).

13 Matthew Schneirov, *The Dream of a New Social Order: Popular Magazines in America 1893–1914* (New York: Columbia University Press, 1994).

14 On the concept of "imagined community," see Benedict Anderson, *Imagined Communities: Reflections on the Origin and Spread of Nationalism*, 2nd ed. (New York: Verso, 1991).

American Exposition in the city of Buffalo, readers across the country shared an explicitly American narrative.[15]

As Louise Appleton notes in her analysis of another iconic national magazine, the *Saturday Evening Post*, it was the "ritual consumption" of the magazines as much as the ideologies they expressed that helped to form a national "imagined community."[16] Hundreds of thousands of American readers were literally and figuratively on the same page. Appleton examines the different levels of scale at which the *Post*, which began its ascent to widespread popularity in 1899, articulated American identity. Through its personalized anecdotal features, short fiction and directive ads, the *Post* brought national and global issues home by addressing readers' experience of the nation at the local and even domestic scales. One's daily behavior during wartime, for example, marked one not so much as a good or bad person but rather as a good or bad American.

In her study of *Reader's Digest*, Joanne Sharp observes similar directive constructions, in this case a project of making the geopolitical personal.[17] Focusing on its treatment of the Soviet Union during the Cold War, Sharp details how the magazine's articles built America's – and Americans' – opposition to the Soviet Union as a moral one. In casting the fear of communism as a moral as much as a geopolitical fear, the state became conflated with individual citizens' values and behavior. As Sharp notes, "The ways in which geopolitical arguments are put forward not only have implications for the ways in which people understand international relations but are also central to the ways in which national identities are formed."[18] *Reader's Digest*, Sharp argues, helped to form a U.S. national identity in opposition to a narrowly construed Soviet identity.

Whether the identity in question is at the individual or national level, the concept of otherness, of difference, is critical in defining a distinct self. Like its popular – and for-profit – magazine compatriots, *National Geographic* helped to articulate a particularly American identity for Americans. This was an American identity in opposition to both old Europe and primitive non-Western regions. It was an identity of civic and technological superiority but yet, a distinctively benign and friendly identity.

15 According to Ohmann (p. 28), the *Ladies' Home Journal* had a circulation of 600,000 by 1891 and according to Schneirov (p. 87), by 1897, Cosmopolitan's circulation was 300,000 and climbing.

16 Louise Appleton, "Distillations of Something Larger: The Local Scale and American National Identity," *Cultural Geographies* 9 (2002), 425. See also Jan Cohn, *Creating America: George Horace Lorimer and the* Saturday Evening Post (Pittsburgh: University of Pittsburgh Press, 1989).

17 Joanne P. Sharp, *Condensing the Cold War:* Reader's Digest *and American Identity* (Minneapolis: University of Minnesota Press, 2000). *Reader's Digest* was started in 1922.

18 Sharp, *Condensing the Cold War*, xvii.

Innocence abroad

Let us return for a moment to the drugstore scene in *It's a Wonderful Life* and remember that the George Bailey who, in a single sentence, links territorial exploration with multitudes of sexually available women, is not the George Bailey immortalized by Jimmy Stewart. This is George as a child.[19] Pronouncing his plans for a couple of harems and three or four wives, young George expresses what he takes to be a man's desires. At the same time, he shows total disregard – literally a deaf ear – for the affections of both Mary and the flirtatious Violet, who had been in the drug store earlier in the scene. His complete lack of interest in girls underscores his youth, and his youthful asexuality disguises his sweeping equation of territorial and sexual control in a veneer of naiveté. Because it is the utterance of a child, this heterosexually suggestive picture of adventure is simultaneously raised and dismissed, functioning as an effective strategy of innocence.

I find the concept of "strategies of innocence," articulated by Mary Louise Pratt in her 1992 book *Imperial Eyes*, which examines European travel writing since the eighteenth century, invaluable in assessing the *National Geographic Magazine* in the period 1888–1945. Pratt defines the term as "strategies of representation whereby European bourgeois subjects seek to secure their innocence in the same moment as they assert European hegemony."[20] Given the unconscious level at which such representations tend to be made, a more appropriate term might be *modes* of innocence, but I like the additional hints of activity and complexity suggested by *strategies*. Although one could argue that "innocence" suggests nonintentionality, often there is intent – that of virtuousness. Improving people's lives through knowledge, through science, through "uplift" are discourses of virtuousness. Still, "innocence" conveys a distancing that I find appropriate for many different aspects of representation in *National Geographic*, and so I examine the strategies of innocence at play in *National Geographic*'s participation in twentieth century American hegemony.

In *Imperial Eyes*, Pratt argues that travel writing was part of the intellectual apparatus of imperialism. While some writers were expressly opposed to European imperial actions abroad, the work of writing travel narratives served to routinize their European bourgeois presence in the "contact zones" and to naturalize European authority. Pratt's overall genre of travel writing is broad, embracing the popular scientific tomes of "the great German geographer" Alexander von Humboldt,[21] as well as the autobiographical adventure tales of Victorian Richard Burton, and

19 Twelve-year-old Bobbie Anderson played young George.

20 Mary Louise Pratt, *Imperial Eyes: Travel Writing and Transculturation* (New York: Routledge, 1992), 7.

21 Preston E. James, *All Possible Worlds: A History of Geographical Ideas* (New York: Bobbs-Merrill, 1972), 147. As James notes, Humboldt "loom[s] large across the pages of the history of science." An active and prolific scientist, who among other accomplishments extensively explored South America in the earliest nineteenth century, Humboldt was also widely read by a lay audience.

the contemporary contemplations of essayist-novelist Joan Didion. Although their eras, projects, nationalities, gender, and "strategies of innocence" differ, all have a geographically based relationship to their subjects, and to their (imagined) readers: They all interpret non-European, non-cosmopolitan, non-bourgeois worlds for a European (or European-derived), cosmopolitan, bourgeois audience.

Some travel writers were specifically interested in providing information for and encouraging European commercial opportunities. Isolating a flurry of such activity in the early- to middle nineteenth century, Pratt refers to a largely British contingent writing about South America as the "capitalist vanguard." But Pratt is more interested in the less aggressively or overtly imperial forms of travel narrative that she calls "anti-conquest" writing, "a utopian, innocent vision of European global authority."[22]

The term *anti-conquest* is both illuminating and troublesome. It is useful in that it attaches seemingly innocent travel narratives to the larger imperial project. The efforts of anti-conquest writers, whether they wished it or not, contributed to European hegemony. But the term is also confusing, because "anti-conquest" narratives were not necessarily *against* conquest. Many writers Pratt describes as anti-conquest seem to have thought that they had nothing to do with European imperialism or commercial concerns. But other anti-conquest writers were perfectly cognizant of, and sometimes even pleased with, their perceived rolein the expanding "European planetary consciousness." While I use the term *anti-conquest*, then, I use it with caution.[23]

What I find most valuable about Pratt's *anti-conquest* formulation for my examination of *National Geographic* is the simultaneous assertion of innocence regarding the larger imperial project and complicity with that project. Under the leadership of Gilbert H. Grosvenor, the Geographic promoted itself as an altruistic educational organization that could be counted on to present solid facts about the world without intrusive politics. And yet, the *National Geographic Magazine* expressed very definite politics, and dressed its facts in fantasies of adventure, travel, sexual availability, Anglo-Saxon superiority, nationalism, and even scientific management. Much as Pratt locates forms of the anti-conquest narrative in different genres, at different times and between different places, I suggest that National Geographic engaged in multiple strategies of innocence.

One of those strategies was science. According to Pratt, science has been one of the most compelling and effective forms of anti-conquest writing, providing an impetus for Western expansion within the presumably innocent framework of pure knowledge and universal human betterment. Pratt goes back to the mid-eighteenth

22 Pratt, *Imperial Eyes*, 39.

23 One contender for a substitute term would be embedded. European and North American writers traveling abroad necessarily were embedded in their own cultural constructs. Like the American journalists embedded in U.S. military units at the start of the Iraq War, embedded travel writers may as individuals possess a critical vision yet their perspective and function are still constrained by the parameters of the dominant geopolitical narrative.

century, when European naturalists began to accept, with enthusiasm, Linnaeus's system of classifying plants and animals. Linnaeus's taxonomic system resembled a global tree of full of blank spaces where a seemingly infinite variety of branches and leaves could be placed. Naturalists journeyed near and far to fill in the blank spaces in the worldwide species map, and travel-writers turned amateur naturalists, eager to contribute to this "global classificatory project."[24] While sometimes even expressly opposed to European imperial actions abroad, these writers served to naturalize scientific and European authority both "home" and "abroad." Establishing a role that the *National Geographic Magazine* would embrace a century or so later, journalists and narrative travel writers became "central agents in legitimating scientific authority."[25]

The development of European science is intertwined with empire-building. Science, and the science derived from exploration in particular, was considered useful knowledge, with its greatest utility in the service of empire.[26] Work on the relationship between geography and imperialism has bloomed in the last decades, engaging multidisciplinary approaches and redirecting the ways in which the histories of geography are told.[27] As Felix Driver maintains, "geography" is not limited to the academic field, but needs to be understood as "an ensemble of practices, institutions and concepts."[28]

In a lovely bit of symmetry, Driver derives the title of his book *Geography Militant* from an essay published in *National Geographic Magazine* in 1924, Joseph

24 Pratt, *Imperial Eyes,* 27.

25 Ibid., 29.

26 See, for example, Robert A. Stafford, *Scientist of Empire: Sir Roderick Murchison, Scientific Exploration and Victorian Imperialism* (Cambridge: Cambridge University Press, 1989) and Felix Driver, *Geography Militant: Cultures of Exploration and Empire* (Oxford: Blackwell, 2001).

27 In addition to the Driver and Stafford books named above, some significant contributions to this literature include Morag Bell, Robin A. Butlin and Michael Heffernan (eds.) *Geography and Imperialism, 1820-1940* (Manchester: Manchester University Press, 1995); Alison Blunt, *Travel, Gender and Imperialism: Mary Kingsley and West Africa* (New York: Guilford Press, 1994); D. Graham Burnett, *Masters of All They Surveyed: Exploration, Geography and a British El Dorado* (Chicago: University of Chicago Press, 2000); Richard Drayton, *Nature's Government: Science, Imperial Britain, and the "Improvement" of the World* (New Haven, CT: Yale University Press, 2000); Felix Driver, "Geography's Empire: Histories of Geographical Knowledge," *Environment and Planning D: Society and Space* 10 (1992): 23–40; Matthew H. Edney, *Mapping an Empire: The Geographic Construction of British India, 1765–1843* (Chicago: University of Chicago Press, 1997); John Gascoigne, *Science in the Service of Empire: Joseph Banks, the British State and the Uses of Science in the Age of Revolution* (New York: Cambridge University Press, 1998); Anne Godlewska and Neil Smith (eds.), *Geography and Empire* (Oxford: Blackwell, 1994); James R. Ryan, *Picturing Empire: Photography and the Visualization of the British Empire* (Chicago: University of Chicago Press, 1997).

28 Driver, *Geography Militant*, 217.

Conrad's "Geography and Some Explorers."[29] In Conrad's nostalgic overview of three phases in the history of geographic knowledge, fantastic tales of other lands were superseded by exploration in search of scientific truths, a heroic period he calls "Geography Militant," exemplified by Captain James Cook and imbued with the romance of adventure and the charge to "fill in" the "blank spaces" on the map. The romance faded with the rise of "Geography Triumphant," the going to where someone (read Westerner) has already gone, the world of modern tourism. Driver notes that *National Geographic* carried on the myth of the heroic explorer solidly into the twentieth century.

As I show in the following chapters, articles in the *National Geographic Magazine* from the late 1880s through the 1940s (and beyond) openly embraced colonial and imperial arrangements, and regularly presumed – and sometimes pronounced – white or Western superiority. A first-page statement in 1898 that the United States should "begin a conquest of the sea no less complete and noble than the conquest of the land already wrought"[30] is not promising evidence for an argument for *National Geographic* as a serialized *anti-conquest* narrative. And yet, as I hope to also show, even these celebrations of conquest are run through with strategies of innocence. The page-one call for conquest proclaims as the necessary goal of American expansion "the ultimate peace and welfare of the world."[31] Even before more complex structures of innocence in the narrative become visible, the violence of imperialism gets overwritten by global "peace." This utopian vision of American expansion, a prevailing position within the Geographic's writers' and editors' milieu, entailed an equation of unencumbered international commerce with harmony among neighbor nations.[32]

The concept of innocence has particular resonance in the twentieth-century American context I explore. Joseph A. Fry suggests that "the image of American innocence and the fundamental uniqueness of imperialism, American style," emerged in full flower in 1898.[33] And Robert Osgood argued that in the wake of the

---

29 Ibid, 3. Joseph Conrad, "Geography and Some Explorers," *National Geographic Magazine* 45 (March 1924): 239–274. Hereafter, in footnote references, I will refer to the *National Geographic Magazine* as *NGM*.

30 W.J. McGee, "The Growth of the United States," *NGM* 9 (September 1898), 386. "No Points" McGee insisted on his name written without periods. See Michael James Lacey, "The Mysteries of Earth-Making Dissolve: A Study of Washington's Intellectual Community and the Origins of American Environmentalism in the Late Nineteenth Century" (PhD diss., George Washington University, 1979); Robert M. Poole, *Explorers House: National Geographic and the World It Made* (New York: Penguin, 2004), 36.

31 McGee, "Growth of the United States," 386.

32 Susan Schulten finds much the same perspective in geography textbooks following the Spanish-American War. See Susan Schulten, *The Geographical Imagination in America, 1880–1950* (Chicago: University of Chicago Press, 2001).

33 Joseph A. Fry, "Imperialism, American Style, 1890–1916," in *American Foreign Relations Reconsidered, 1890-1993*, ed. Gordon Martel (New York: Routledge, 1994), 67. See also Howard J. Wiarda, *Cracks in the Consensus: Debating the Democracy Agenda in U.S.*

rhetoric-laden Spanish-American War, "although the nation had tasted the first fruits of world power, most Americans lingered in the age of innocence, naive and scornful witnesses of the unsentimental calculation of national advantage which preoccupied the minds of military men and power-conscious statesmen and scholars."[34] I suggest that American lingering in the "age of innocence" was not due to nostalgic attachment but rather to the fact that the dominant discourse would not have it any other way.

The Chilean writer Ariel Dorfman suggests that there is something particular to the history of the United States, and the timing and methods of its empire-building, that has allowed it to be

> interpreted, time and again, as the domain of innocence. In a sense, a more extraordinary feat than changing thirteen colonies into a global empire in less than two centuries is that the U.S. managed to do it without its people losing their basic intuition that they were good, clean, and wholesome. … They desired the power which can only come from being large, aggressive, and overbearing; but simultaneously only felt comfortable if other people assented to the image they had of themselves as naive, frolicsome, unable to harm a mouse.[35]

Dorfman finds it significant that United States' rise to global preeminence coincides with the development of mass communications, providing more forms and more outlets for mythologizing the country's history and the future. National Geographic, which actively developed its reputation as a wholesome educational source and whose accepted role as scientific authority bolstered its claim to political disinterest, was one of the more important outlets.

Setting scales of analysis

Presenting America's World has a bifurcated structure. What I aim to do in the following five chapters is to provide a dual scale of analysis: *National Geographic* as a singular identifiable entity engaged in a particular cultural landscape, and *National Geographic* as a composite, however tightly managed, of individual cultural producers with their own agendas.

Chapter 1 sets the National Geographic Society and its magazine within the intellectual, cultural and political contexts of the late nineteenth century. Established as a journal largely by and for scientists, *National Geographic Magazine*'s earliest mastheads were a roll call of well-placed government scientists, working

---

Foreign Policy (Westport, CT: Praeger, 1997) and Jim Hanson, *The Decline of the American Empire* (Westport, CT: Praeger, 1993).

34 Robert Endicott Osgood, *Ideals and Self-Interest in America's Foreign Relations: The Great Transformation of the Twentieth Century* (Chicago: University of Chicago Press, 1953), 57.

35 Ariel Dorfman, *The Empire's Old Clothes: What the Lone Ranger, Babar, and Other Innocent Heroes Do to Our Minds* (New York: Penguin Books, 1983), 201–2.

in departments and agencies such as the United States Geological Survey, the Bureau of the Census, and the Department of Agriculture. The chapter examines the early history of the National Geographic Society within the contexts of the professionalization of the sciences, the development of popular magazines, and American overseas expansion.

The 1898 Spanish-American War was a critical turning point. The sudden attention in the United States to the rest of the world, and the U.S.'s new imperial position, shook all strata of American geography, just at the time that efforts were being made to solidify geography as a profession and a scholarly pursuit. The National Geographic Society became caught in the struggle over definitions of geography and the structure of the discipline. Some of this struggle was played out in the internal politics of the Society and the magazine, and culminated in the establishment of what is today the official organization in the United States for academic geographers, the Association of American Geographers. As for the Geographic, as Susan Schulten notes, "it was the Spanish-American War that first suggested to the editors that geography included the realm of human – even political – interaction."[36]

The Geographic's popular approach toward "humanized geography" drew such a response that by 1919, National Geographic promotional literature proclaimed the Society to be "by far the largest scientific body in the world."[37] Giving geography a broad definition – "the world itself and all it holds,"[38] National Geographic similarly defined its general science in terms of facts, exploration, and discovery. Funds accrued through increasing membership and advertising were funneled into expeditions that in turn both legitimized the Society as a scientific organization and provided original material for the magazine.

"The world and all that is in it" had a different ring to it when the nation to which the Society belonged had a pronounced political hand halfway across the globe. The Spanish-American War led to many changes within the Geographic, within the discipline of geography, and to the nation as a whole. Even after a history of continual continental expansion, Americans still needed to interpret this new world of United States overseas imperialism. It was Alexander Graham Bell, who became president of the National Geographic Society in January 1898, who saw the opportunity for the magazine to provide "scientifically reliable" information in a manner more like that of the popular *Century* magazine.[39] Bell, who succeeded the Society's inaugural president – and his father-in-law – Gardiner Greene Hubbard, hired Gilbert H. Grosvenor to become the magazine's first paid editor. Grosvenor married Bell's daughter Elsie soon thereafter.

36 Schulten, *Geographical Imagination in America*, 67.

37 "The National Geographic Magazine: Specimens in Miniature of What It Has Given in the Past and Will Bring in the Future," pamphlet, 1919: 2, file 2, Item XVI, Box 2, Maynard Owen Williams Collection, Kalamazoo College, Kalamazoo, Michigan [hereafter Williams Collection].

38 Alexander Graham Bell [hereafter AGB] to Gilbert Hovey Grosvenor [hereafter GHG], 4 April 1904, Box 267, Bell Papers.

39 AGB to GHG, 13 July 1899, Box 99, Grosvenor Papers.

Chapter 2 probes many of the layers of National Geographic relationships, including those with the nation its title represented. With its close ties to federal officials as high as the President and its advocacy of federal policies, the National Geographic Society enjoyed its ambiguous status as an independent entity with governmental-weight authority. As Grosvenor noted in 1916, "The National Geographic Society is generally regarded as semi-official."[40] The Society, "historically a darling of Capital Hill," was also a tax-exempt nonprofit organization, with modifications to its tax status initiated only in 1994.[41]

Founded in the midst of the late nineteenth century current of American positivism, *National Geographic* came to represent a conservative brand of Progressivism. In this chapter I examine the dimensions of the political beliefs embraced by the Geographic from its formation through its twentieth-century leadership by Gilbert H. Grosvenor. Despite clear political rhetoric and editorial decisions, *National Geographic* promotional literature and internal doctrine emphasized the magazine's avoidance of politics. "The success of the magazine must always rest," declared Grosvenor in 1919, on "retaining the confidence of the public in the absolute accuracy and impartiality of its contents."[42] The principle of nonpartisanship was designed to protect perceptions of objectivity and purity of purpose, and to create a timelessness that would permit the *National Geographic Magazine* to function as a sort of installment encyclopedia of random places, animals, and adventures. And as the success of the Geographic attests, the approach worked. But this assertive denial of politics, while backed by a select avoidance of political discussion – a decision with political import of its own – was basically rhetorical; I find it to be one of the National Geographic's strategies of innocence.

Chapter 3 looks at how scientific order was used to put people in their place. I use "look at" deliberately, since this chapter focuses on the visual portrayal of peoples and places within the "global classificatory project." It is "the primal act of witnessing," Steven Greenblatt has said, "around which the entire discourse of travel is constructed."[43] By the twentieth century, photography had became the favored recording of such witnessing. Building on Pratt, Anne McClintock suggests that photography provided the classification project with an important tool of "European planetary consciousness." Concurring that "the imperial science of the surface promised to unroll over the earth a single 'Great Map of Mankind,' and cast a single, European, male authority over the whole of the planet," McClintock suggests that "the promoters of the global project

40 GHG to AGB, 26 October 1916, Box 100, Grosvenor Papers.

41 Constance L. Hays, "Seeing Green in a Yellow Border: Quest for Profits is Shaking a Quiet Realm," *The New York Times*, 3 August 1997, sec. 3:1, 12.

42 Report of the Director and Editor of the National Geographic Society, 1919, Box 160, Grosvenor Papers. As Schulten notes, this particular assertion may have been in response to stagnating National Geographic membership during World War I, when the magazine ran many articles clearly in favor of Britain and against Germany even before the U.S. officially picked a side.

43 Stephen Greenblatt, *Marvelous Possessions: The Wonder of the New World* (Chicago: University of Chicago Press, 1991), 122.

sorely lacked the technical capacity to formally reproduce the optical 'truth' of nature as well as the economic capacity to distribute this truth for global consumption" until the later nineteenth century and the emergence "of commodity spectacle – in particular photography."[44]

Photography brought a technical accessibility to "true reality" that sketches, prints and paintings could not, turning visual representation into a form of reproduction. "The bright colors of tropic climes and picturesque native figures ... [could now be] reproduced exactly as they appear to the actual beholder in their homelands."[45] Some of the earliest books of photographs and photographed-derived engravings, which were easier to print, were of places quite distant from the European metropoli in which they were published. As if to demonstrate photography's role as a tool of empire, the "first photographically illustrated travel book" featured photographs, commissioned by the French government, of French-controlled territories in North Africa and the Middle East.[46] As James Ryan notes, photography provided "an indispensable record of the progress and achievement of Empire."[47]

Photography was the cornerstone of the *National Geographic Magazine*'s success. "People want to *see* things," Gilbert Grosvenor told *The American Magazine* in 1922. "The most expressive words, marshaled by a master of language, would not convey the bizarre ferocity of an African savage in his war garb. But a good photo of the gentleman, scowling into the camera, gives almost as vivid an impression as a face-to-face encounter."[48] Initiated as "An Illustrated Monthly" in 1901, *National Geographic*'s devotion to photography solidified after a "bold" decision in 1905 to publish 11 pages of photographs of Lhasa delivered compliments to the editor – and a boost in membership. By 1915, *National Geographic Magazine* had established its own photographic laboratory. It was the first publication to use a number of different photographic processes, particularly in color film.

National Geographic photographs catered to a certain range of aesthetics, such as the iconic ethnography familiar both to viewers of postcards and stereographs as well as to readers of anthropological studies. The *Geographic* prided itself on pictures that were beautiful, colorful, and classically aesthetic. At the same time, however, *National Geographic* advertised its photographs as scientific documentation, delivering fact in pictorial form. "Fact and fancy, exercise of the reason and of the imagination, training towards the logical and towards the visionary, surely these are pairs implying not merely antithesis but antagonism," argued Martin Johnson in

---

44 Anne McClintock, *Imperial Leather: Race, Gender and Sexuality in the Colonial Context* (New York: Routledge, 1995), 34.

45 "Specimens in Miniature," 10.

46 The book is Maxime Du Camp's 1852 *Égypte, Nubie, Palestine et Syrie*. Melissa Banta and Curtis M. Hinsley, *From Site to Sight: Anthropology, Photography, and the Power of Imagery* (Cambridge, MA: Peabody Museum Press, 1986), 39.

47 Ryan, *Picturing Empire*, 11.

48 Allison Gray, "We All Have a Secret Love of Adventure and Romance," *The American Magazine* (May 1922), 27. We shall see, however, in Chapter 4, that not all staffers believed that a scowling subject made for a good *National Geographic* photograph.

1949. But he also saw a similar aim to both art and science: communication. "Their legitimacy in each case," he noted, "lies in the power to evoke coherent mental imagery."[49] Chapter 3 discusses how the modern entwining of art and science acted to facilitate communication between scientists and the larger public and how it served the interests of imperial governments.[50] Particularly in its photography, *National Geographic* employed an amalgam of science and art to communicate pictures of the world. In the context of the *National Geographic Magazine*, science and art functioned as two complementary, if often contradictory, strategies of innocence.

I consider how romance and aesthetics serve as companionate narrative structures of innocence. Critics have decried aestheticization, particularly in the practice of photography of colonial places and people, a "cover-up of evil under the sign of beauty and rarity."[51] In *National Geographic*'s world, political context was anathema to beauty because it sullied the pure forms of natural and cultural landscapes – people included – with a temporal reality. Combining the discourses of art/aesthetics and science, "clouds of fantasy and pellets of information,"[52] photography proved the perfect medium for *National Geographic*'s anti-conquest narrative.

One form of representation common to both science and art, and prevalent in travel and exploration narratives, is the "type." Nineteenth-century scientists, in particular, were quick to adopt photography in their efforts to discern and define human types. These efforts, in turn, fueled colonial rule: directly by assisting government identification and management of indigenous people, and indirectly by providing the stamps of detachment and fact onto pictures broadly disseminated within the culture of the imperial powers. While the "type" was not the only photographic convention appearing in the *National Geographic Magazine*, I focus on it because of the powerful way it combined both scientific and aesthetic discourses within photography, especially in the photography of the world's people.

Perhaps the most talked about photographic genre in the *National Geographic Magazine* was that of the erotic figure. The trope of the "porno-tropics," as Anne McClintock puts it, was familiar in European and American narratives well before *National Geographic* began publishing photographs of scantily clad, dark-skinned, healthy young women and men.[53] The erotic exotic has strong psycho-metaphorical as well as enacted political manifestations. Tiffany and Adams call the former the "romance of the wild woman," in which the remote and unknown woman/land

---

49 Martin Johnson, *Art and Scientific Thought: Historical Studies Towards a Modern Revision of Their Antagonism* (New York: AMS Press, 1949), 23, 192.

50 By *modern* I mean from the Enlightenment onward.

51 W.J.T. Mitchell, *Picture Theory: Essays on Verbal and Visual Representation* (Chicago: University of Chicago Press, 1994), 309, discussing Malek Alloula, *The Colonial Harem* (Minneapolis: University of Minnesota Press, 1986).

52 Susan Sontag, *On Photography* (New York: Farrar, Straus and Giroux, 1973), 63.

53 McClintock, *Imperial Leather*, 21. See also Bernard Smith, *European Vision and the South Pacific* (Oxford University Press, 1960).

beckons to the man, promising pleasure and fertility to he who is able to tame and control the temptress and her unpredictable dangers.[54]

But the Western male eroticization of real, rather then metaphorical, "other" women, has also had much to do with European class designations, and the sexualities ascribed to them by middle-class society. In much of the nineteenth and well into the twentieth century, no one, and certainly not the "well-bred" man, was to think of "well-bred" women as sexual creatures. Refined white women were the achievement of civilization, and thus the farthest of anyone from an animal nature. Working-class women, however, were another story. Indeed, "working girl" as a euphemism for prostitute betrays the conflation of class with available sexuality. Colonized or non-Western women were at least as far from the European feminine bourgeois as the European working-class, and probably more so.[55] In practice, the marketed sexual desire of the colonizing forces served as an assertion of power over both the men and women of the colonized region.[56] Sexuality is an inextricable element of "othering," and of race. While the meaning and categories of "race" have varied significantly over the last three hundred years, the concept is imbued with biological concepts, prominently among these, heredity. The existence of racial designations at all has much to do with fears of those in power regarding the sexuality and procreation of those not in power, those designated as racially different.[57]

In both Chapters 3 and 4 I discuss the notable presence of dark, bare-breasted women in *National Geographic*. More often than I would have expected, word association in casual conversation with non-geographers led them from geography to *National Geographic*, and from *National Geographic* to bare-breasted women (that from men in particular). The question of how the subject of geography was two degrees away from photographs of nearly naked women was what launched

54 Sharon W. Tiffany and Kathleen J. Adams, *The Wild Women: An Inquiry into the Anthropology of an Idea* (Cambridge, MA: Schenkman, 1985). On the erotic feminization of American landscapes, see also Annette Kolodny, *The Lay of the Land: Metaphor as Experience and History in American Life and Letters* (Chapel Hill: University of North Carolina Press, 1975) and Henry Nash Smith, *Virgin Land: The American West as Symbol and Myth* (Cambridge: Harvard University Press, 1950).

55 See Sander L. Gilman, "Black Bodies, White Bodies: Toward an Iconography of Female Sexuality in Late Nineteenth Century Art, Medicine, and Literature," *Critical Inquiry* 12 (1985): 204–42; Joanna de Groot, "'Sex' and 'Race': the Construction of Language and Image in the Nineteenth Century," in *Sexuality and Subordination*, eds. Susan Mendus and Jane Randall (London: Routledge, 1989); Sarah Graham-Brown, *Images of Women: The Portrayal of Women in Photography of the Middle East 1860–1950* (New York: Columbia University Press, 1988).

56 See Alloula, *Colonial Harem*; Ann Stoler, "Making Empire Respectable: the Politics of Race and Sexual Morality in 20th Century Colonial Cultures," *American Ethnologist* 16 (1989): 634–60; Vron Ware, *Beyond the Pale: White Women, Racism and History* (London: Verso, 1992); Robert J.C. Young, *Colonial Desire: Hybridity in Theory, Culture and Race* (London and New York: Routledge, 1995).

57 McClintock, *Imperial Leather*; Gilman, "Black Bodies"; Nancy Stepan, "Race and Gender: The Role of Analogy in Science," *Isis* 77 (June 1985), 261–277.

me on this project, interrogating the intertwining of gender, race, and sexuality in the construction and representation of *National Geographic*'s world. [58] Chapter 3 discusses how the magazine drew on both aesthetic and scientific rationales for its photographs of lightly attired women and men, constructing complementary strategies of innocence that allowed for "just that seemingly odd juxtaposition of stodgy respectability and half-naked women."[59] Chapter 4 examines the experience of one *National Geographic* photographer, Maynard Owen Williams, in this form of representation.

The contributors

Hegemony, of course, is not fixed. It is, rather, a process, a continuing effort to win consent. *National Geographic* as a case study reveals several scales at which the manufacturing of consent works. At one level is *National Geographic* as an active participant in the developing United States hegemony within the modern capitalist world system. A second level, closely intertwined with the first, engages the magazine's readers and deals with the world that the magazine presented to its largely American audience. A third level at which I examine the manufacturing of consent, and perhaps a less obvious one, is within the institutional world of employment at *National Geographic*.

A magazine is, with some small exceptions, a collaborative effort.[60] Most commercial magazines have a publisher, an editor in chief, an illustrations editor, an editor in charge of the layout of the magazine, and varying ranks of other editors. In addition, there are also the writers and illustrators who contribute either as freelancers or as part of the staff. Having left a career in journalism to pursue graduate work in geography and zigzagged between journalism and academia since then, I have a considerable interest in the editorial process, particularly in the dance of politics and personality between editor and writer, "boss" and "worker." It is not unusual for journalists' political beliefs to be at odds with those of their publisher, if not of their highest-ranking editors. Selective assignments, self-censorship and editorial changes are the most frequent ways in which publishers, editors and writers negotiate their politics. In this internal arena of manufacturing consent, the writer either works within the system, gently pushing the bounds of acceptability, or leaves.

"As a protection of the Society and Magazine, it has been the policy of your editor to collect around him a staff of efficient men," reported Grosvenor in 1919,

---

58 Tamar Rothenberg, 'Voyeurs of Imperialism: *The National Geographic Magazine* Before World War II,' In *Geography and Empire*, eds. Anne Godlewska and Neil Smith (Oxford: Basil Blackwell, 1994), 155–172. Catherine Lutz and Jane Collins cover similar ground, extending into the second half of the twentieth century. Catherine A. Lutz and Jane L. Collins, *Reading National Geographic* (Chicago: University of Chicago Press, 1993).

59 Rothenberg, "Voyeurs," 156.

60 "Zines," which tend to be small in pages and distribution, are frequently one person's project.

> and I think I may confidently state that no magazine today possesses as competent, broad-minded and enthusiastic a corps, all trained for their specialties, as the Geographic. Many of our staff have been repeatedly invited elsewhere, but the ideals of the National Geographic Society have bound them so strongly that they would not leave.[61]

Grosvenor's use of martial analogy in describing his staff was likely inspired by the recently concluded World War. Yet this statement also reveals a pronounced hierarchy with a clear Chief, and a familiarly nationalistic style blurring of identity between the collective organization (the National Geographic Society), the abstract ideals and goals associated with the organization, and the organization's leader.

For more than 50 years, Grosvenor was the gatekeeper at the Geographic. As he and the magazine matured, Grosvenor solidified his control over the magazine and the National Geographic Society, and the magazine's politics became very much his own. Chapter 2 examines Grosvenor's politics and their expression in the magazine's editorial policy. Chapters 4 and 5 present two different contributors to *National Geographic*. Maynard Owen Williams, the subject of Chapter 4, joined the *National Geographic* staff as a writer and photographer in 1919, and produced or contributed to nearly 100 articles for the magazine before his retirement in 1960. Harriet Chalmers Adams, the subject of Chapter 5, was not a staff member, but contributed 21 articles to *National Geographic* between 1907 and 1935, wrote several National Geographic Society news bulletins for newspaper syndication, and was a regular speaker at the Society's lecture series.

I have chosen to focus on Williams and Adams because they – or rather, their archival materials – provide a perspective on National Geographic that is both internal and external in its relationship to the magazine and the Society. Williams was a dedicated Geographic loyalist, and a staff member for 40 years, but he frequently felt like an outsider. This status was in part self-imposed, as Williams disliked working at the National Geographic home office in Washington, D.C., and maneuvered as much as possible for overseas assignments. But Williams also consistently drew criticism from his editors; they berated him for writing over the heads of the readers. Later in Williams's career, editors considered both his writing and photography somewhat outdated and so hesitated in proffering article assignments. Adams, on the other hand, seems to have been as much a Geographic insider as anyone without ever holding an official position. When not traveling, she based herself in Washington, D.C., and counted the Grosvenors among her social circle. Letters between these two contributors and their *National Geographic* editors, including Grosvenor, managing editor John Oliver La Gorce, and illustrations editor Franklin L. Fisher, detail some of the editorial processes of the magazine.

I also examine the lives of Williams and Adams to interrogate manifestations of gender in the world in which they lived, particularly in the realm of travel, and in National Geographic's world. Williams was an affable white Midwestern and middle-class man who was experienced in both the missionary and journalist

61 "Report of the Director and Editor of the National Geographic Society," 1919, Box 160, Grosvenor Papers.

traditions of travel and travel writing before he arrived at *National Geographic*. In many ways, his was the life that the fictional George Bailey coveted, a life of adventure and women and seeing the world. Writing home from Mendan, in what is now Indonesia, Williams noted that he was "ordering 120 dozen more films – 1440 more exposures. More baggage, more traipsing about seeking for [sic] the right view point, more cajoling natives into cooperation. It's a swell game."[62] What Williams wanted most from his photographic subjects was a good smile and colorful outfits, but his musings on the subject of bare brown breasts reveal much about his thoughts on sexuality and social meaning.

Harriet Chalmers Adams began her exploring career in the early years of the twentieth century, when rough travel and geographical or economic expertise were strongly identified with men and masculinity. During her lifetime, judging from newspaper articles, personal letters and memoranda, Adams was known for a particular combination of characteristics: courageous explorer, Latin America expert, charming lady. Adams's femininity was as important to her audience – including the press – as was her subject matter. While not the only woman to be accorded authority in the masculine-coded arenas of exploration, geography and commerce, the fact of Adams's gender was unusual enough for it to be a constant issue. Still, her femaleness conformed to class expectations – she was "ladylike" rather than strongly sexual, for example, more Katherine Hepburn than Mae West.

Yet Adams was forced to be conscious of her gender in ways that male explorers or travelers or scientists or writers or lecturers were not. She was one of the early members and first president of the Society of Women Geographers, formed in part because the all-male Explorers Club refused to accept women as members. Although professing indifference to the question of women's suffrage prior to women's winning the vote in 1920, Adams made it clear through her statements and actions that she believed a woman's place was wherever woman wanted to be. She took notice of women's activities and status in the places she visited and seems to have been particularly inspired by South American feminists.[63]

Adams was also an active participant in the economic expansion of the United States, working with the Pan American Union, lecturing on trade with Latin America, and encouraging U.S. investment in South America. Her initial travels in South America were entwined with her husband's work on behalf of a mining company and her *National Geographic Magazine* articles expressed positive excitement over the advancement of American and European commerce in places previously on the periphery of global capitalism.

Williams, on the other hand, had increasing doubts about the beneficence of global capitalism and of missionary Christianity, and he was apprehensive about his abetting of American tourism. With his consciousness of and conflicting

62 Maynard Owen Williams [hereafter MOW] to Daisy Woods Williams [hereafter DWW], 15 June 1937, Item XX, Box 3, Williams Collection.

63 M. Kathryn Davis, "The Forgotten Life of Harriet Chalmers Adams: Geographer, Explorer, Feminist" (Masters Thesis, San Francisco State University, 1995).

responses to his role as participant in the American version of "European planetary consciousness," Williams offers a case study in the contradictions of the anti-conquest position, particularly within the context of American imperialism.

On a methodological note, I focus on Williams and Adams because, of all the possible frequent contributors to *National Geographic* during the first half of the twentieth century, they are the only ones whose work is accessible to scholars. Unfortunately, in an ironic counterpoint to the National Geographic Society's stated mission to "diffuse geographic knowledge," the vast wealth of archival material at the National Geographic Society is generally inaccessible to scholars and journalists. No outsider is allowed free rein in the meticulously organized Records Library, although it is possible for a researcher to access a file by knowing very specific information, such as the existence of a letter from person X to person Y on a particular day, or the name and dates of a particular assignment. Access to these records, many of which have been microfiched, would allow for the creation of a fuller historical account of exactly how geographic ideas were brought to the public.

Other readers, other stories

The difficulty of access to the National Geographic Society archives has discouraged scholarly or investigative work on the society and its magazine. On the occasion of its centennial, the National Geographic Society published a congratulatory coffee-table tome by C.D.B. Bryan replete with the photographs that made the magazine so popular. Bryan, who as the official historian for this purpose was able to use the records library, discusses a few unflattering aspects of the magazine's history, but keeps the revelations light enough so that evidence of racism, misogyny and fascism is reduced more to individual foibles rather than to institutional predilection.[64] His book builds on and supersedes the smaller official histories written by Gilbert H. Grosvenor and published in varying forms as a small book published by the society and as an essay in the 1888–1946 magazine index.[65] Bryan's affectionate history is matched by journalist Howard Abramson's more critical 1987 history, *National Geographic: Behind America's Lens on the World*, which exposes the magazine's right-leaning politics, takes the Geographic to task for its support of polar explorer Robert E. Peary over rival explorer Frederick Cook, and interrogates the society's long-held status as a non-profit, tax-exempt organization.[66] A later, gentler work by

64 Bryan, *100 Years*.

65 Gilbert Grosvenor, *The National Geographic Society and Its Magazine* (Washington, D.C.: National Geographic Society, 1957); Gilbert Hovey Grosvenor, "The Story of the Geographic," in *National Geographic Index 1888-1946* (Washington, D.C.: National Geographic Society, 1967).

66 Howard S. Abramson, *National Geographic: Behind America's Lens on the World* (New York: Crown Publishers, 1987). Abramson also published *Hero in Disgrace: The True Discoverer of the North Pole, Frederick A. Cook* (iUniverse, 2000), excoriating the National Geographic Society's role in championing Peary as discoverer over Cook.

former *National Geographic* editor Robert M. Poole tells the story of the Geographic through a focus on the Hubbard-Bell-Grosvenor family. An engaging narrative of personalities, Poole's 2004 *Explorers House* also reveals the racism and anti-Semitism of the Geographic's leadership, while embracing the magazine's approach to adventure, photography and the natural world.[67]

One of the important cornerstones of the Geographic's reputation was its image as a scientific organization, the veracity of the magazine's information derived not just from journalistic integrity but from scientific objectivity. Historian Philip Pauly's 1979 *American Quarterly* article focuses on the development of the Geographic from its origins as one of many scientific societies established in the nation's capital in the late nineteenth century.[68] These societies served as centers for the consolidation of disciplinary and professional information and many, like the National Geographic Society, defined their mission as the "diffusion of knowledge" beyond their disciplinary circles as well. Natural history museums embraced the mantle of popularizing science, presenting three-dimensional narratives of nature and "man's" place in – and in control of – nature. Donna Haraway, in her essay "Teddy Bear Patriarchy," examines the ideologies of nature, imperialist politics, nationalism and masculinity at work at the American Museum of Natural History, paralleling the same messages purveyed by the *Geographic*.[69] Lisa Bloom, in her 1993 *Gender on Ice*, delves further into issues of nationalism and masculinity, particularly as framed by *National Geographic*.[70] She focuses on the Geographic's support for and portrayal of Peary's polar explorations as emblematic of the magazine's "discourse of nationalism, empire and white male heroism" and shows how the Geographic used Peary to cement its blended reputation as authoritative source for serious science, medium for tales of heroism, and quasi-official national representative.[71]

Bloom is one of a handful of scholars who must have thought, following National Geographic's centennial, "Where are the critiques?" That's what prompted me to examine gender and race in the magazine's photographs.[72] Julie Tuason turned her attention to the ideology of empire in the magazine's plentiful coverage following the Spanish-American War. Tuason's 1999 *Geographical Review* article notes the

67 Poole, *Explorers House*, 2004.

68 Philip J. Pauly, "The World and All That Is In It: The National Geographic Society, 1888-1918," *American Quarterly* 31 (1979): 517–32.

69 Reprinted in a number of forms and forums, "Teddy Bear Patriarchy: Taxidermy in the Garden of Eden, New York City, 1908–1936" appears as the second chapter of Donna Haraway's *Primate Visions: Gender, Race and Nature in the World of Modern Science* (New York: Routledge, 1989). *Primate Visions* chapter "Apes in Eden, Apes in Space" critiques more recent (post-1950s) National Geographic constructions of gender, race and nature.

70 Lisa Bloom, *Gender on Ice: American Ideologies of Polar Expeditions* (Minneapolis: University of Minnesota Press, 1993). Upon re-reading the book in 2006, I realized that Bloom also seized on the scene in *It's a Wonderful Life* as emblematic of the Geographic's racialized and sexualized place in American male adolescent fantasies (73, 75).

71 Bloom, *Gender on Ice*, 58.

72 Rothenberg, "Voyeurs of Imperialism," 1994.

consistency of the magazine's positive outlook on the United States' control of the Philippines while the overall narrative shifted from a largely commercial focus immediately after the war to an emphasis on broader economic development and "moral tutelage."[73] Linda Steet examines the interplay of colonialism, imperialism and Orientalism of *National Geographic*'s portrayal of Arabs, finding a remarkable consistency. Over its first 100 years, the magazine devoted a disproportionate number of articles on the Middle East and North Africa, and Steet shows the *Geographic* to be a reliable purveyor of Orientalism. Both Maynard Owen Williams and Harriet Chalmers Adams come in for criticism; they fit the overall *Geographic* perspective on the Arab world.[74]

Catherine Lutz and Jane Collins' 1993 *Reading National Geographic*, like my own examination, stems from political concerns "about the imaginative spaces that non-Western peoples occupy and the tropes and stories that organize their existence in Western minds."[75] They too had been struck by the dearth of criticism of *National Geographic*, especially given the magazine's prominence in American culture. More interested in contemporary content and readings of the magazine, Lutz and Collins examine *National Geographic* from the 1950s onward, mixing interpretative methods familiar in literary criticism with quantitative techniques, interviews with *Geographic* staff, and controlled-study interviews of "ordinary people."

As an example of *National Geographic*'s role as a cultural marker, Lutz and Collins point to a 1990 Ralph Lauren advertisement, reprinted in the book.[76] Stacks of the magazine, foregrounded in an interior hunting lodge scene, help to indicate "good taste and upper-middle-class lifestyle." Importantly, however, the magazines displayed in this 1990 advertisement have the oak-leaved, text-only covers that predate the 1960s. As such, the magazines stand not only for good taste but for continuity of class position and longstanding American nationality.

My interest is in the origins and early development of *National Geographic* as a cultural standard-bearer. While Lutz and Collins do cover the origins and growth of the National Geographic Society and magazine, their main concern is the magazine's already established place of prominence in what they call an "American-fashioned world."[77] For example, they examine *National Geographic*'s photographs of non-Western women against the backdrop of media portrayals of American women in the 1950s, 1960s, and 1970s – periods of ideological debate over women's place in the home and in waged labor. Contextualizing such representations within the post-war "cult of domesticity," Lutz and Collins examine the magazine's pictorial odes to the

73 Julie A. Tuason, "The Ideology of Empire in National Geographic Magazine's Coverage of the Philippines, 1889–1908," *The Geographical Review* 89 (January 1999), 35.

74 Linda Steet, *Veils and Daggers: A Century of* National Geographic*'s Representation of the Arab World* (Philadelphia: Temple University Press, 2000). The book is based on her dissertation, "Teaching Orientalism in American Popular Education: National Geographic, 1888–1988" (PhD diss., Syracuse University, 1993).

75 Lutz and Collins, *Reading National Geographic*, 2.

76 Ibid., 9.

77 Ibid., xiii.

housewife-mother, as well as its presentation of young non-Western women as both feminine and laborers. While their analysis is fruitful, because they ignore decades of *National Geographic* representations of women, it is incomplete. Images of woman as mother and non-Western women as worker were longstanding staples of *National Geographic*, going all the way back to the early twentieth century.

Gillian Rose uses Lutz and Collins' quantitative examination of postwar *National Geographic* photographs as an example of a content analysis methodology.[78] While I think that technique can offer interesting insights, I do not believe it necessarily captures what lingers in someone's reading or looking at a publication. Memory selects, certain ideas and images stand out. What launched me on this project was other people's specific association of bare-breasted women with *National Geographic*. Throughout its existence, the magazine has portrayed a vast range of people and landscapes, as well as animals and plants. Nude or nearly-nude natives appear only sporadically and sometimes readers would have to wait several issues to get their next peek. If *National Geographic* mostly ran photographs of bare-breasted women, it is unlikely that it would have achieved the mass popularity and virtuous reputation that it did. And yet, it is these selective images that stand out and leave an impact on readers. As Steet notes, a racist statement is not balanced out by a non-racist statement, nor does it make sense to imagine a formula for how many clothed women it takes to cancel out a bare-breasted woman.[79] For this project, then, I looked for the representations that stood out to me. I looked through at least one issue in every year of the magazine from 1888 to 1945, looking especially for articles on what would come to be called the Third World, most of which at the time were places under colonial rule. I then also looked at the other articles in my selected issue.

By beginning their examination of *National Geographic* after both the magazine and American hegemony in the world system had been established, Lutz and Collins chose not to dwell on how the magazine actually developed itself in direct relation to American overseas expansion. In looking at the way in which *National Geographic* has represented the "American-fashioned world," they skipped the years in which that world was forged, and the way in which *National Geographic* fashioned itself in the service of this process. One of the aims of this book is to understand the connections between American expansion and *National Geographic*'s support of that enterprise.

In some respects, Susan Schulten's *The Geographical Imagination in America, 1880–1950* already beat me to it. Schulten's study, though, examines not just National Geographic but geographical representations in school textbooks and academia as well as in popular commercial cartography. She demonstrates the paradigmatic upheaval the Spanish-American War caused in popular and scholarly geography, and the continuing adjustments made in each field as the United States' position in world affairs changed. Schulten's discussion of how changing cartographic techniques and technologies, some developed by National Geographic, helped shape the way

78 Gillian Rose, *Visual Methodologies* (London: Sage Publications, 2001).

79 Steet, *Veils and Daggers*, 10.

Americans literally saw the world, adds an important dimension to the discussion of the interpretation of images perceived as scientific facts.[80]

Through an examination of readers' letters, Schulten achieves an understanding of the give and take between a magazine and its audience. As any editor knows, you can't please everybody all the time, and neither do you want to, but most editors at most popular publications have a vested interest in providing the material that readers want. Clearly, the world that *National Geographic* presented was a world that its readers – a million of them by the mid-1920s – wanted to know about.

These days our ideas about people and places in the world derive from a multitude of sources – school, family, friends and acquaintances, religious institutions and a dizzying variety of mass media. But for much of the first half of the twentieth century, it was *National Geographic* that provided Americans – middle-class white Americans, in particular – with an ongoing narrative of the world and how to see it. A beautifully presented 1915 pamphlet declared that "the Geographic Magazine removes the padlock of technical terms from the portals of geographic science and invites the world to share its delights."[81] And indeed, whether they were devoted members of the National Geographic Society or came across the magazine in classrooms or doctors' offices, millions of people accepted the Geographic's invitation.

Histories of geography have tended to start off wide and general: "nature and culture in western thought," general ideas and inclinations, theories by great geographical thinkers from Ptolemy and Strabo to Immanuel Kant. Then they follow geography into the institutional walls of the universities, where it is isolated from the larger, "general," culture.[82] Of course, geography – loosely, the study of the earth, and of spatially conceptualized relationships between and among places and people on earth – is a cultural product itself. Academic trends, professional geographical practices and presentations of popular geography all reflect and participate in larger cultural shifts and concerns. And recently, geographers, historians and kindred spirits have addressed the meaning and uses of geography in the larger culture.[83] "What is

80 Schulten, *Geographical Imagination in America*. The book is based on her dissertation, "The Transformation of World Geography in American Life, 1880–1950" (PhD diss., University of Pennsylvania, 1995).

81 La Gorce, *Story of the Geographic.*

82 Clarence J. Glacken, *Traces on the Rhodian Shore: Nature and Culture in Western Thought from Ancient Times to the End of the Eighteenth Century* (Berkeley: University of California Press, 1967); David N. Livingstone, *The Geographical Tradition: Episodes in the History of a Contested Enterprise* (Oxford: Blackwell, 1992). Livingstone's book, as he takes pains to note, is structured around important intellectual moments in geography rather than as a linear progression, but the effect of wide base to narrow academic one is similar to that found in Glacken. The most prominent adherent of the "great geographer" approach is James, *All Possible Worlds*.

83 Driver and Blunt, for example, are geographers, Schulten is a historian, Lutz and Collins are anthropologists and Bloom has a varied cultural studies background.

needed," as Felix Driver wrote in 1992, "is greater attention to the ways in which geographical knowledge is presented, represented, and misrepresented."[84]

Attending Driver's exhortation, I will be using "strategies of innocence" as a point of entry into the complex world view of National Geographic. Taking cues from ongoing European narratives of benevolent colonialism, National Geographic provided American versions of imperialistic altruism and uplift that negated the violence and arrogance of imperialism, and even negated the existence of a political perspective at all. Under the guise of objective science, the magazine gave readers familiar tropes of racial distinction, the primitive, the Arab, the sexually available woman, and so on. Both science and art worked together to allow for the "innocent" display of bare-breasted women, impure thoughts displaced through truth and beauty.

National Geographic used these strategies in its text, its photography and its self-representation, strategies that continue to have implications today. They have their echo in what has become the pervasive American projection of innocence in the face of imperialist activities, a posture that is perhaps only now beginning to crumble. By examining some of these powerful, if subtle, rhetorical strategies, and the ways in which individual contributors to *National Geographic* situated themselves in relation to both the magazine and to American and European imperialism, I hope to contribute to a serious treatment of popular geography.

84 Driver, 'Geography's Empire,' p. 35.

Chapter 1

National Geographic in the New World Order

Founded in 1888, the National Geographic Society was a testimony to the excitement in the nation's capital in the late nineteenth century regarding both the professionalization of science and its popularization. By the turn of the century, however, professionalization and popularization were no longer considered compatible within a single organization. Geography, with its new and tenuous existence in American research universities, was particularly sensitive to this tension – or at least, academic geographers were. After a few years of internal struggle, the National Geographic Society emerged, through its magazine, as the face of popular geography; scholarly geographers formed their own professional organization.

Part of a general shift towards professionalization and the formation of vocational associations (as opposed to the more avocational societies of amateur practitioners), American scientific societies proliferated in the late nineteenth century. Societies for economics, entomology, chemistry, physiology, geology, and biology were founded in the 1880s; within the following decade, the fields of mathematics, physics, astronomy, and psychology all established parallel associations. These were based in Washington, D.C., where the bulk of their active members worked for various federal agencies.[1] "As the nation grew, inquiries concerning resources and the conditions of material development became necessary," recounted National Geographic Society vice-president W J McGee in 1898, leading to the establishment of many federal scientific offices.[2] The United States Geological Survey, Department of Agriculture, Bureau of the Census, Bureau of American Ethnology, Coast and Geodetic Survey, and Weather Bureau were among the federal agencies drawing scientists and the scientifically-minded to Washington D.C. "In time the experts voluntarily met for mutual benefit and grouped themselves in unofficial organizations, which now stand

---

1 Pauly, "World and All"; Lacey, "Earth-Making Dissolve"; James K. Flack, *Desideratum in Washington: The Intellectual Community in the Capital City, 1870–1900* (Cambridge, MA: Schenkman, 1975). The American Anthropological Association was founded in 1902. Among disciplines outside the sciences, the Modern Language Association formed in 1883, the American Historical Association in 1884, the American Philosophical Association in 1900 and the American Political Science Association in 1903.

2 W.J. McGee, "American Geographic Education," *National Geographic Magazine* 9 (July 1898), 306.

in the front rank of learned societies of the world; and official bureaus and unofficial societies are one in purpose, and that the highest in human reach – the increase and diffusion of knowledge for human weal," McGee explained.[3]

McGee believed that "ideally the scientific society, embracing educated and intelligent laymen as well as experts, should counteract specialization by diffusing knowledge in reasonably sophisticated form."[4] Increasing specialization and the refinement of disciplines as universities began to expand their research departments were manifestations of the professionalization of science and scholarly work. This would shortly become a delicate issue for geography. The division of the sciences into separately focused societies did not mean that individuals were expected to limit their interests or research to a single defined subject area. McGee, for example, was active in at least 11 other scientific societies based in Washington D.C., and could as easily have called himself a geomorphologist as an ethnologist.[5] As much as these societies were designed for the use of professional practitioners, their memberships were generally open to anyone – as the Entomological Society expressed it in its constitution – in any way interested" in a society's purpose.[6] McGee considered such organizations the "intermediate link" between scientific specialists and the general public.[7]

The National Geographic Society could also count itself among the various geographical societies in Europe and the Americas, some founded as early as the 1820s and 1830s. These societies existed as centers of geographical information, broadly construed to include commercial, botanical, geological and anthropological angles, among others, with emphasis on knowledge about and derived from exploration. Membership, too, drew from an assortment of vocations and interests, attracting university professors, diplomats, gentleman scientists, colonial administrators, army and naval officers, and amateur explorers, among others. Driver describes Britain's influential Royal Geographic Society (RGS) as "part social club, part learned society, part imperial information exchange and part platform for the promotion

---

3 Ibid. Many of the same scientists belonged to the Cosmos Club, a social organization and building for the scientific elite established in Washington in 1878.

4 Curtis M. Hinsley, Jr. *Savages and Scientists: The Smithsonian Institution and the Development of American Anthropology 1846–1910* (Washington, D.C.: Smithsonian Institution Press, 1981), 234.

5 The path from geology to ethnography is not an unfamiliar one in the history of geography during the late nineteenth and early twentieth centuries. See Livingstone, *Geographical Tradition.* Other "polyspecialist" contemporaries of McGee include Charles Merriam, Lester Ward, John Wesley Powell, and Henry Gannett (see Lacey, "Earth-Making Dissolve"). Gannett, who was Chief Geographer of the USGS for 32 years, was a topographic cartographer as well as a census statistician. He was president of the National Geographic Society from 1910 until his death in 1914 (Lacey, "Earth-Making Dissolve"; Bryan, *100 Years*).

6 Flack, *Desideratum in Washington*, 144.

7 Hinsley, *Savages and Scientists*, 234.

of sensational feats of exploration," particularly of Africa.[8] The RGS emphasized geography as a practical, rather than a theoretical discipline, and promoted it as "the science of empire."[9] Geographic societies facilitated and encouraged nationalist projects of empire-building, which in turn sustained the societies and gave them purpose.

In the Americas, where the New World was still being explored, Mexico City and Rio de Janeiro established geographical societies in the 1830s. It was not until 1851 that the first such organization in the United States, the American Geographical Society (AGS), was founded in New York City. Inspired by the RGS, its founders "conceived the chief purpose of geography to be the empirical description and mapping of the earth. This necessitated exploring expeditions to fill in the blank spaces… ."[10] Members consisted of significant numbers of businessmen, who sought new markets, accessible resources and safe and efficient passage to and from these areas, as well as newspaper editors, writers, and publishers, for whom expeditions made good news items. The American Geographical Society also attracted professional geographers, and it developed a notable book and map library. Its *Bulletin* concentrated on exploration, as well as featuring articles on disasters and major engineering projects. Interestingly, the National Geographic Society had a greater influence on its New York counterpart than the American Geographic Society had on the Washington upstart; in 1895 the AGS shifted its focus from exploration to the development of geography as a professional and educational discipline. The *Bulletin* started running more articles on physical geography and other subjects deemed more useful for the teaching of geography.[11]

The National Geographic Society's mission, as reported in the first issue of the magazine, in October 1888, was "to increase and diffuse geographic knowledge," with the magazine one means towards that end.[12] The concept of "increasing and diffusing knowledge," had been embraced by "self-culture" organizations in both Britain and the United States from the 1820s through the 1840s. These early industrial-era societies, such as the Society for the Diffusion of Useful Knowledge (British, founded 1825) and its namesake and emulator, the Boston Society for the Diffusion of Useful Knowledge (founded 1829) were among several institutes organized by philanthropic elites to provide opportunities for moral and intellectual improvement for skilled workers such as clerks, mechanics and artisans. These societies established libraries and offered public lectures on such topics as biology, Christianity, slavery and pneumatics.[13]

8 Driver, *Geography Militant*, 25.

9 Ibid., 27.

10 John Kirtland Wright, *Geography in the Making: The American Geographical Society 1851–1951* (New York: American Geographical Society, 1952), 12.

11 Wright, *Geography in the Making*.

12 "Announcement," *NGM* 1 (October 1888), i.

13 William L. Bird, Jr., "A Suggestion Concerning James Smithson's Concept of 'Increase and Diffusion,'" *Technology and Culture* 24 (April 1983): 246–255; Howard M. Wach, "'Expansive Intellect and Moral Agency': Public Culture in Antebellum Boston," *Proceedings of the Massachusetts Historical Society* 107 (1995): 30–56; Howard M. Wach,

More immediately, the National Geographic Society's declaration paid homage to the Smithsonian Institution, with which some of its members were actively affiliated. Based in Washington, D.C., the Smithsonian was founded in 1840 with a $500,000 bequest by James Smithson to the United States "for the increase and diffusion of Knowledge among men."[14]

The knowledge to be increased and diffused by the new Society would be specifically, if also generally, geographical: "As it is hoped to diffuse as well as increase knowledge, due prominence will be given to the educational aspect of geographic matters, and efforts will be made to stimulate an interest in original sources of information."[15] Articles on geomorphology ran alongside detailed expositions of such occurrences as the Great Blizzard of 1888 and the planning for the Nicaragua Canal. Publication was sporadic, ranging from quarterly to nearly monthly to twice a year, until the magazine was established as a monthly in 1896. Into the first few years of the twentieth century, the *National Geographic Magazine* carried articles by eminent academic geographers such as William Morris Davis, Ralph Tarr, and Mark Jefferson as well as reports from officials in the various federal offices, especially branches of the United States Geological Survey.

The reign of the *National Geographic Magazine* as a journal of choice for professional geographers was relatively brief, and coincided with an emphasis in school geography on the subject as a physical science. Harvard professor William Morris Davis, "the Dean of American Geographers,"[16] successfully convinced the National Education Association to make physical geography a college entrance requirement in 1893, focusing and legitimizing the discipline around a rigorous, evolutionary science core.[17] Previously, geography textbooks had offered a broader range of interpretations of the subject, alternately giving geography political, cultural or physical foci.[18] But the political impact of the events of 1898 would soon overshadow the new emphasis on physical science in textbooks and the *National Geographic Magazine.*

1898: Turning point for the nation and the *National Geographic*

For many in geography – and at the *National Geographic Magazine* – the real excitement began in 1898. "It is doubtful if the study of any branch of human

---

"Culture and the Middle Classes: Popular Knowledge in Industrial Manchester," *Journal of British Studies* 27 (October 1988): 375–404.

14 Bird, Jr., "Smithson's Concept," 246; Hinsley, *Savages and Scientists,* 17. Lacey, "Earth-Making Dissolve," and Pauly, "World and All," both suggest that the Smithsonian was an important model for the National Geographic Society.

15 "Announcement," i.

16 Thomas F. Barton and P.P. Karan, *Leaders in American Geography, vol. 1: Geographic Education* (Mesilla, NM: New Mexico Geographical Society, 1992), 25.

17 See Schulten, *Geographical Imagination in America.*

18 Ibid.

knowledge ever before received so sudden and powerful a stimulus as the events of the past year have given to the study of geography," declared John Hyde, editor of the *National Geographic Magazine*, in the June 1899 issue. In April 1898, the United States went to war with Spain, fought in and over Spain's colonial possessions in North America and Asia. The U.S. won the war in a matter of weeks, taking over Cuba, Puerto Rico, and the Philippines. In the aftermath, the United States formally annexed the islands of Hawaii, where American expatriates had overthrown the Hawaiian monarchy five years earlier.[19] By the end of 1898, the United States had expanded beyond the continent and acquired an overseas empire.[20]

In addition to the actual acquisition of islands and archipelagoes in the Pacific and Atlantic, the Spanish-American War was significant for its rhetoric. The explanation for the war with Spain – to help the Cubans liberate themselves from Spanish rule – built itself on a resuscitated concept of the United States as torchbearer for revolutionary democratic freedom. The torch this time, however, was lit by newspapers; "there seems to be great probability in the frequently reiterated statement that if Hearst had not challenged Pulitzer to a circulation contest at the time of the Cuban insurrection, there would have been no Spanish-American War."[21] Two leading New York newspapers, Hearst's *Journal* and Pulitzer's *World*, each trying to outdo the other in sensationalism and scoops, helped whip up their readers' feelings against Spain and for the Cuban revolutionaries.[22] When the U.S. battleship *Maine* exploded in Havana in February 1898, the *Journal* and the *World* led the pack crying for outright U.S. war against Spain, despite lack of evidence that Spain had anything to do with the *Maine*'s destruction. Although there was substantial opposition to the war and to the United States' hold on Spain's former colonies in the war's wake, the yellow press succeeded in amassing popular support for the war.[23]

---

19 The successful 1893 coup leaders had called for annexation to the United States, but the U.S. president at that time, Grover Cleveland, displeased with the dirty maneuvering, responded that the United States "did not go around overthrowing foreign governments," and declined the annexation. (H.W. Brands, *Bound to Empire: The United States and the Philippines* (New York: Oxford University Press, 1992), 17.

20 The U.S. also claimed the Pacific islands of Guam, Wake, Midway, Howland, Baker, and Samoa (some of them) in 1898.

21 Frank Luther Mott, *American Journalism: A History: 1690-1960*, 3rd ed. (New York: Macmillan, 1962), 527.

22 Both the New York *World* and the New York *Journal* sold their news services, including pictures, to newspapers all over the U.S., and so spread the national frenzy over Cuba. In addition, national syndicates with their own pools of reporters operated in the late 1890s. See Mott, *American Journalism*, and also Marcus M. Wilkerson, *Public Opinion and the Spanish-American War: A Study in War Propaganda* (New York: Russell & Russell, 1967 [1932]).

23 See, for example, Brands, *Bound to Empire*; Daniel B. Schirmer, *Republic or Empire: American Resistance to the Philippine War* (Cambridge, MA: Schenkman, 1972); Michael H. Hunt, *Ideology and United States Foreign Policy* (New Haven: Yale University Press, 1987); Julius W. Pratt, *The Expansionists of 1898: The Acquisition of Hawaii and the Spanish Islands* (Gloucester, MA: Peter Smith, 1959). The term "yellow press" to describe sensationalist papers

Geographers, including professional and academic geographers, textbook authors and the cartographic industry, were among those who responded favorably to the 1898 turn of events.[24] The board of the National Geographic Society was no exception, and at a special meeting at the end of that year, granted U.S. President William McKinley honorary membership. The committee formed to notify McKinley of this honor explained to him "that it was the design of the Society to signalize the beneficent changes of the year in the modification of the civil geography of the world, and that the action was to be understood as an endorsement by one of the leading scientific organizations of the country of the course of the President as a great national leader."[25]

The 1898 acquisitions were portrayed in the magazine as a positive and natural step in the spread of U.S. enlightened democracy and enterprise, a glorious expression of manifest destiny. "America must soon lead the world in ocean navigation as in other directions, and begin a conquest of the sea no less complete and noble than the conquest of the land already wrought," declared NGS vice president McGee in an address delivered to the joint session of the National Geographic Society and the American Association for the Advancement of Science in August 1898, and published as the leading article in the September 1898 issue of the *National Geographic Magazine*. "More than anything else, the territorial acquisitions must contribute toward the extension of enlightenment, toward the elevation of humanity, and toward the ultimate peace and welfare of the world."[26]

McGee developed his thesis of United States progressive expansion further, and in June 1899 the magazine published a related piece, again as the lead article. Here, McGee invokes Science to explain the naturalness and goodness of U.S. expansion, bringing his hyperbolic style to bear on the anti-imperialist feeling to which he was apparently responding:

> Today the trembling ones shrink shrieking at the self-conjured ghost of imperialism, as if empire could grow in freedom's soil, as if the bright-winged papilio of constitutional law might, forsooth, creep back to the chrysalis where the monarchial pupa grew in centuries past; ... they ignore the Law of Human Progress (seen through the coordination of other sciences in the Science of Man) under which humanity moves, in ways *orderly as planetary orbits* of vital stages, from savagery into barbarism, thence into civilization, and finally into enlightenment, never dropping backward save by extinction; they comprehend not the full significance of humanity's law, vaguely expressed as "manifest destiny," which proves that imperialism is impossible on the plane of enlightenment, and that the peoples

---

such as the *World* and the *Journal* was derived from a popular comic that ran in both papers, the "Yellow Kid." (Mott, *American Journalism*, 525–526).

24 Schulten, *Geographical Imagination in America.*

25 "Proceedings of the National Geographic Society, Session 1898–99," *NGM* 10 (April 1899), 143. The meeting took place on December 9, 1898. Before the war, the magazine had published an article by Henry Gannett arguing against U.S. annexation of overseas territory because the process would incorporate a less advanced population, to the detriment of the nation as a whole. Gannett, "The Annexation Fever," *NGM* 8 (December 1897): 354–358.

26 McGee, "Growth of the United States," 386.

of the earth are steadily rising from plane to plane with the certainty of ultimate union on the highest of the series. [italics my emphasis][27]

McGee takes great pains to explain the expansionism of the United States as progressive, meaning both that it followed steps in a logical, incremental series, and that each step was an improvement over the last. McGee's statement that "humanity's law … *proves* that imperialism is impossible on the plane of enlightenment," [my emphasis] reflected the concept of the scientific nature of social and political phenomena, although it is hard to resist interpreting his argument as a Lady Macbeth effort, protesting too much. Earlier in the piece McGee had argued that U.S. expansion had been humanitarian, not pecuniary (with the annexation of Alaska the one exception), and that the overseas expansion of 1898 was clearly in the same vein as the nation's prior growth spurts. His denial of U.S. expansionism as imperialism was meant to differentiate the U.S. experience from the European experience. He characterized the United States as more modern than Europe, better endowed with both natural resources and enterprising souls, and healthier in every way. This view of the United States in comparison with Europe was a familiar one in both progressive and conservative (including "progressive conservative") circles at the time; it would continue to dominate *National Geographic*'s coverage of the United States and of Europe for at least the next 20 years.[28]

McGee may have downplayed the commercial motives of the Spanish-American War and its subsequent territorial boon for the United States, but the *National Geographic Magazine* certainly acknowledged the commercial and financial benefits of having the additional territory, with articles evaluating the economic conditions in the Philippines and Samoa, as well as articles recounting the economic benefit of colonies to European countries.[29] Given the high percentage of the *Geographic*'s board associated with the federal government's scientific departments, it is not surprising that the magazine supported the United States' overseas activities. As Schulten notes, "a strong national presence internationally would strengthen their fields, just as a firmly grounded science might enhance the nation's position abroad."[30] Lacking the pretense of neutrality that the magazine would later promote, the *National Geographic Magazine* of this period functioned both as a scientific journal and an outlet for articulating the government's perspective.[31] For example, O.P. Austin, who was chief of the U.S. Bureau of Statistics at the turn of the century,

27 W.J. McGee, "National Growth and National Character," *NGM* 10 (June 1899), 204.

28 See Cohn, *Creating America*. The *Saturday Evening Post*, under the editorship of George Horace Lorimer, made this distinction even more forcefully and less generously. Philip Pauly discusses *National Geographic Magazine*'s portrayal of modern America vs. quaint Europe in an unpublished paper, "The National Geographic Society and the Iconography of an Emerging World," Johns Hopkins University, 1976.

29 See Tuason, "Ideology of Empire."

30 Schulten, *Geographical Imagination in America*, 54.

31 Ibid., 46.

contributed articles to the Geographic based on the government reports he produced in his official capacity.[32]

Among the most "beneficent changes in the modification of civil geography" for the National Geographic Society was that the sudden acquisition of unfamiliar and distant territories provided lots of opportunities for newly relevant material. The magazine published a full issue on Cuba in May 1898, and full issues on the Philippines in June 1898 and February 1899.[33] As Gilbert H. Grosvenor told the story in retrospect, "American interest in world affairs was awakened by the Spanish-American War," and the *National Geographic Magazine* turned its focus towards that popular interest.[34]

The acquisition of new territories also provided a major impetus for the National Geographic Society's decision to hire a full-time editorial staff member. The Society's president since early 1898, Alexander Graham Bell, recommended "the adoption of the policy of national expansion," explaining that "it has been my policy since taking the reigns of office to enlarge our outside membership so as to place the Society on a national plane." Most of the National Geographic members lived in Washington, D.C. or its vicinity; those members who did not relied on the magazine for their connection to the Society since they could not attend meetings or lectures. Bell determined that expansion beyond the capital area would require an appealing publication that came out on a regular basis, and so the Society should hire a "salaried officer ... to act as Assistant Editor, whose sole duty it should be to attend to the Magazine and foster the growth of the outside members."[35] The editorial committee until then had consisted of volunteers such as McGee, Gannett, and John Hyde, most of whom had full-time government jobs.

Bell had been elected Society president after the death of its first president, Gardiner Greene Hubbard, in December 1897. Unlike the majority of the National

---

32 Austin's "Colonial Systems of the World," *NGM* 10 (January 1899): 21–26, for example, derived from a Bureau of Statistics report entitled "Summary of Commerce and Finance," December 1898, uses the examples of Britain and France to explain the commercial benefits to the United States of its possessing Cuba, Hawaii, Puerto Rico, and especially, the Philippines. Austin later (1904–1932) held the post of Secretary of the National Geographic Society.

33 Over the following ten years, the *National Geographic Magazine* published at least 11 articles on Cuba, 24 articles on the Philippines – including another full issue in May 1898 – and ten articles on Puerto Rico (the magazine firmly defended that spelling, rather than Porto Rico). Hawaii during this period did not get very much attention, with specific articles only in the December 1904 and April 1908 issues. Samoa got some early attention, with articles in June 1899 ("The Commercial Importance of Samoa" and "Samoa: Navigators Islands") and November 1900 ("The Samoan Islands"), and the magazine published an article about Guam in May 1905 ("Our Smallest Possession – Guam").

34 Grosvenor, *NGS and Its Magazine*, 7.

35 Alexander Graham Bell, "Address of the President of the National Geographic Society to the Board of Managers," 1 June 1900, Box 31, A. W. Greely Papers, Library of Congress. Published in *NGM* 11 (October 1899): 401–408.

Geographic Society's active membership and board of managers for its first ten years, Hubbard was not a scientist. He was, however, a wealthy and educated supporter of scientific research. A regent of the Smithsonian Institution, Hubbard backed such projects as the journal *Science*, which eventually became the organ of the American Association for the Advancement of Science. While his investment in *Science* never paid off – he ended up selling the unprofitable journal for $25 – his investment in technology proved highly rewarding.[36] It was he who funded Bell's experiments in telephony, and when Bell got his patent, it was Hubbard who established the Bell Telephone Company.

Bell's first relationship with Hubbard had been as tutor to Hubbard's daughter Mabel. Mabel had lost her hearing in a childhood bout with scarlet fever, and Bell, following in the footsteps of his father, was an innovative teacher of the deaf. In the late 1860s, Alexander Graham Bell and his parents emigrated from Scotland to Canada. Bell met Hubbard in Boston during a lecture tour demonstrating his father's technique for teaching speech to the deaf. Hubbard hired Bell to tutor his daughter. Bell fell in love with Mabel, which she eventually reciprocated, and after Bell secured his future by inventing the telephone, they married.[37]

It was Bell the father as much as Bell the president of the National Geographic Society who wrote to Edwin A. Grosvenor, professor of European history at Amherst College, to see if either of his twin sons, both of whom had graduated at the top of their class at Amherst, would be interested in the position of assistant editor and assistant secretary of the National Geographic Society, at a salary of $1200 a year. Although Bell phrased his letter as if he were inquiring at all the best colleges and universities, and as if there was no difference between the twins, he knew that his daughter Elsie had some interest in Gilbert Grosvenor, who, with his brother Edwin P. Grosvenor, had been summer houseguests of the Bells for a few weeks two years earlier. To no one's surprise, young Edwin decided to continue pursuing his law studies. Gilbert, known as Bert, who was teaching at a private school in New Jersey, quit his job before the semester was over and moved to Washington.[38] His salary, in an arrangement Bell had worked out earlier with the board, came out of Bell's pocket rather than from the coffers of National Geographic.

The National Geographic Society was faltering financially when Bell took the helm in January 1898, but Bell was determined to ensure the National Geographic Society's existence as a useful organization. Useful meant reaching as many people as possible. And in the wake of the Spanish-American War, it became clear that not only did Americans need information about places in the world, they *wanted* that information. The *National Geographic Magazine*, already the Society's broadest means of outreach, would become the focus of the Society. "From the very first it has been a really valuable journal, but limited in its sphere of usefulness to the members of

36 Grosvenor, *NGS and Its Magazine*, 22.

37 Lacey, "Earth-Making Dissolve"; Grosvenor, *NGS and Its Magazine*; Bryan, *100 Years*; Poole, *Explorers House*.

38 Grosvenor, *NGS and Its Magazine*, 19–23; Poole, *Explorers House*.

the National Geographic Society," explained Bell. "We desire now, through its means, to benefit the whole country."[39]

"Having in my mind as an ideal a magazine like the *Century*, of popular interest and yet scientifically reliable as a source of Geographic information,"[40] Bell turned to the experts for advice, sending out nearly identical letters to the editors and publishers of the top magazines of the day. "We are anxious to give a wider circulation to the National Geographic Magazine than it has yet received," Bell wrote to Richard Watson Gilder, editor of the *Century*, and to S.S. McClure, publisher of *McClure's*.[41] Gilder had guided the *Century* to its position in the mid-1880s as the "leading American quality magazine," known for its history and biography articles, travel stories, essays on social reform, short fiction, and high quality woodcut illustrations.[42] McClure used the *Century* as a model when he launched his own monthly magazine in 1893, but pushed the format toward a broader audience by chopping the cost of subscriptions and cultivating lively "behind the scenes" articles.[43]

The Geographic's magazine, with its scientific origins, geographical focus, and circulation of about 1,000,[44] posed no competition to the popular literary and current event magazines, and Bell engaged in fruitful conversations with the other magazine editors. One suggestion, from a former editor of the *Atlantic Monthly*, that the National Geographic Society should reach the schools and enliven geographic learning, pleased Bell immensely. "We must have today in the United States at least thirty millions of persons under the age of twenty-one years. Just think what an enormous field of usefulness there would be for our magazine if it could reach teachers."[45]

To attract more readers, the magazine needed to make itself bigger and brighter. "But we can't enlarge the magazine without more financial support," Bell told Grosvenor, "which can only come from an increase in the number of corresponding members and from advertisements."[46] Without the circulation, the magazine would not draw advertisers, and by the turn of the century, it was quite clear that advertising

39 AGB to Richard Watson Gilder, 16 August 1899, Box 267, Bell Papers.

40 AGB to GHG, 13 July 1899, Box 99, Grosvenor Papers.

41 AGB to Richard Watson Gilder, 16 August 1899, Box 267, Bell Papers; AGB to S. S. McClure, 16 August 1899, Box 267, Bell Papers. The letter noted that "The Society has unrivaled opportunities for obtaining geographic news of the greatest interest to the public generally through the various departments of the Federal Government; and we have upon our editorial staff prominent representatives of these departments who give us inside news. We aim to make the National Geographic Magazine a reservoir of information to which the press of the country can resort for reliable and timely items relating to all the geographic topics that may be occupying the public mind."

42 Schneirov, *New Social Order*, 45.

43 Ibid, 80; Ohmann, *Selling Culture*.

44 AGB, "Chart Showing Increase in Membership of the National Geographic Society," Box 160, Grosvenor Papers.

45 AGB to GHG, 28 September 1899, Box 267, Bell Papers.

46 Ibid.

could more than sustain a publication. The *Century*, Bell's exemplar, had been the first "dignified magazine edited for a dignified reader group" to seek out advertising, in the 1870s and 1880s.[47] Then in the 1890s, *McClure's* and *Munsey's*, aiming for a larger, more middle-class audience, led the way in dropping subscription prices and using advertising to cover the majority of publication costs.[48]

Both potential members and potential advertisers had to be made aware of the magazine. Grosvenor was charged with the task of creating positive publicity about the magazine and securing its reputation. This involved persuading other magazines to review, refer to, or reprint excerpts of items in the *National Geographic Magazine*. To build membership, Grosvenor also recruited people individually. As promotional pamphlets from the early twentieth century onward emphasized, one *joined* the National Geographic Society, becoming a member who supported the work of the Society – including its expeditions – and received the magazine as a benefit of membership. One needed to be nominated to the Society in order to join; until the 1950s, the signature of a sponsoring member was necessary, if easily obtainable.[49] The arrangement successfully created an imagined community of National Geographic Society members, all invisibly connected not merely through common readership, but through membership in a single organization.

That community rested on class identification. Under Bell's aegis, the membership base that Grosvenor initially sought consisted of the wealthy, the connected, and the respected. Grosvenor began by sending out letters of nomination to the National Geographic Society to "prominent names" compiled by Bell from membership lists of organizations such as the National Education Association and the National Academy of the Sciences. "You better send out these copies marked personal," Bell told Grosvenor, "with some indication that they are sent by me."[50] Bell also had Grosvenor publish an account of a hunting expedition in Greenland that both provided the magazine with the first-hand adventure-style writing that would become its hallmark and roped in a new bunch of "young men with means" for membership.[51] "I think it would pay to give him space and then get the young men of that hunting party to nominate corresponding members; they all, I understand, belong to important families and must have many friends who would be glad to see their names mentioned in the magazine," said

---

47 Theodore Peterson, *Magazines in the Twentieth Century*, 2nd ed. (Urbana: University of Illinois Press, 1964), 21.

48 Ohmann, *Selling Culture*; Schneirov, *New World Order*.

49 Abramson, *Behind America's Lens*, 153; Ishbel Ross, "Geography, Inc.," *Scribner's Magazine* 103 (June 1938), 24. According to Ross, "If an applicant lives in a small town and cannot find a sponsor, the [membership] committee finds one for him." Those residing in prison were deemed undeserving of membership, but for an extra 50 cents a year could receive the magazine as subscribers only. The notorious Al Capone, for example, was stripped of his membership when he was sent to prison, and put on the subscription list instead. (The story may well be apocryphal.)

50 AGB to GHG, 16 August 1899, Box 267, Bell Papers.

51 AGB to GHG, 21 September 1899, Box 267, Bell Papers.

Bell.[52] Fullerton Merrill's article, "A Hunting Trip to Northern Greenland," his sole contribution to the magazine, appeared in the March 1900 issue.

Beyond individual appeals, Grosvenor's task was "to make the magazine as attractive and interesting as possible to the general public, for this will intend to increase its circulation."[53] Bell saw cultivation of elites as an important first step, but he had in mind a much broader readership for the magazine, women included. Considering the potential for an audience of schoolchildren, Bell wrote to Grosvenor in 1899 that, "Wherever there is anything of interest going on we should have geographic information to illustrate it. This is not only good doctrine for small boys, but for grown men and women too."[54] The magazine would take a more informal – although no less "accurate" – approach to imparting geographic facts to a larger audience. "Dry and longwinded articles of technical character should be avoided as much as possible," Bell told Grosvenor. The revised magazine should offer readers "a multitude of good illustrations and maps."[55] In addition to examining the popular magazines of the day, Grosvenor studied what might be considered geographical best sellers, from Herodotus's *History* to Darwin's *Voyage of the Beagle*; as he recounted later, "each was an accurate, eyewitness, firsthand account. Each contained simple, straightforward writing – writing that sought to make pictures in the reader's mind."[56]

Openly visible pictures – actual photographs – were the most important element of the revamped magazine. "[T]he first requisite is accuracy and reliability of statement," wrote Bell; "the second which is no less important and indeed is absolutely essential to success is that the matter should be presented in an interesting and entertaining way – in language free from technicality known only to the expert." The third requirement "is that the magazine should be plentifully illustrated with pictures of living interest – pictures of the dynamical rather than the statical variety."[57] As Bell later commented, "Judging others by myself I should say that the features of the most interest are the illustrations and the little foot-notes that accompany them."[58] Grosvenor was already familiar with the attraction of published photographs of people and places; his father's book on Constantinople, published in 1895, was "the first scholarly work to be profusely illustrated by photoengravings."[59] Bell would later assure his son-in-law that, "In my opinion the magazine has been SAVED by the pictures, and by the fact that you have the ability to select illustrations of general interest."[60]

52 AGB to GHG, 18 October 1899, Box 267, Bell Papers.

53 AGB to GHG, typewritten version 13 July 1899, Box 99, Grosvenor Papers.

54 AGB to GHG, 28 September 1899, Box 267, Bell Papers.

55 Ibid.

56 Quoted in Bryan, *100 Years*, 42.

57 AGB to GHG, 7 December 1905, Box 267, Bell Papers.

58 AGB, 23 October 1907, Home Notes, Box 99, Grosvenor Papers.

59 Grosvenor, *NGS and Its Magazine*, 21.

60 AGB to GHG, 4 April 1904, Box 267, Bell Papers. In his 1957 retrospective, Grosvenor cites an 1896 *National Geographic Magazine* article by W.J. McGee to demonstrate

By the end of Grosvenor's first year at the Geographic, the Society's membership had doubled, from 1,000 to 2,200.[61] But Bell's enthusiasm for the magazine's new direction and its new director was not necessarily shared by those on the editorial committee. Apparently, a dispute over the spelling of some place names in July 1900 led the Executive Committee of the magazine to nullify Grosvenor's yearly contract; he could be dismissed with a month's notice. His position as assistant editor was downgraded. His function was now spelled out as "acting as executive officer of the Editorial Committee and of the Editor in chief, *carrying out their instructions,* in the correction and editing of manuscripts for publication, in the arrangements for illustrations, and in the reading and correction of proofs."[62] President Bell, conveniently in Europe at the time of the committee's action, returned immediately, called together the Board of Managers, and solidified Grosvenor's position, promoting him to managing editor and increasing his salary.[63]

The next "coup attempt" was led by the magazine's editor-in-chief, John Hyde. Hyde, who volunteered his time, was less pleased by the move towards popularization and perhaps resentful that he was being replaced as director of the magazine by a paid neophyte.[64] He took the opportunity of Grosvenor's wedding in London to Elsie Bell in August 1900 to undermine the young man's position at the magazine. Leveling "charges of ignorance, incapacity and laziness" against Grosvenor,[65]

---

the divide between the earlier magazine and its popular revision. "[T]he excessive use of picture and anecdote is discouraged," wrote McGee; "superficial description and pictorial illustration shall be subordinate to the exposition of relations and principles." The distinction in direction is striking here, but would have been more so had McGee actually been referring to the magazine, as Grosvenor implies (Grosvenor, *NGS and Its Magazine*, 39). McGee's context is not the magazine, however, but the lecture series. (W J McGee, "The Work of the National Geographic Society," *NGM* 7 [August 1896], 258). At the time Grosvenor was writing the book in 1957, however, he was suffering from a "terrific acute sinus attack." GHG to Maynard Owen Williams, 23 March 1961, File "Grosvenor Bio Notes," Item 29, Box 7, Williams Collection.

61 AGB, "Chart Showing Increase in Membership of the National Geographic Society," Box 160, Grosvenor Papers.

62 Henry Gannett, Chairman of the Executive Committee, to GHG, 27 July 1900, Box 99, Grosvenor Papers. My italics.

63 Grosvenor, *NGS and Its Magazine*, 32.

64 Writing to his father in early June, 1900, Grosvenor told him: "I can't help feeling that Mr. Hyde wants to retain the editorship after all, and thought that if he could get someone else in my place, he would be sure to stay in. I do not like to suspect him, but when he tells Mr. Bell that he loves me like a son, and with the same breath that I am lacking in business ability and weak in proof-reading and hence do not deserve any increase in salary (not in these words, but to that effect), it is funny to say the least." (Bryan, *100 Years*, 48).

65 "In my opinion the opposition to him centers in Mr Hyde who for some reason or other appears desirous of driving him out of the Magazine," Gilbert Grosvenor's father wrote upon hearing of Gilbert's troubles with the board. "Whether one be my son or not, I have an intense sympathy with a young man. It makes my blood boil when from some of those above him a young man receives the very opposite of appreciation after he has striven to do his work

Hyde persuaded the Board of Managers, at least some of whom were rankled by their magazine's shift away from solid scientific fare and towards geography-as-travel-and-adventure, to hand the publishing of the magazine over to S. S. McClure, with the ubiquitous American News Corporation handling newsstand sales. If the *National Geographic Magazine* were to be a popular magazine, they reasoned, then it should be handled like a popular magazine, published out of New York by a proven magazine publisher and sold on newsstands like other magazines. The magazine, possibly with a new name, could maintain a nominal connection to the Society and produce income, while the Society would continue its sturdy record of publication in a more professional journal.[66]

Grosvenor cut short his European honeymoon to try to rescue his job and his magazine. Perhaps to offset any possible accusations of nepotism, Bell absented himself from the dispute, leaving it up to Grosvenor to fight for himself. Grosvenor was eventually able to convince the Board of Managers that not only was it actually much cheaper to publish the magazine out of Washington, but that he had the ability and determination to make the magazine a financial success of high repute. By the end of 1901, the Society distinguished between its popular and its technical meetings (the latter held at the private Cosmos Club, home away from home to the scientific elite of Washington), and the magazine – with Henry Gannett, not John Hyde, as editor-in-chief – carried the subtitle "An Illustrated Monthly."[67] McClure continued to publish the magazine through the April 1903 issue, the second issue under Grosvenor as editor-in-chief.

The move toward popularization also disturbed some of the professional geographers who had considered the National Geographic Society an institutional home base and its magazine a scientific journal. Harvard professor William Morris Davis had "several of his more famous papers" published in the pre-Grosvenor magazine,[68] and was on the Society's Board of Managers when he was asked to join the editorial Executive Committee in 1902. He declined, saying he "couldn't approve general policy of popularization at expense of science."[69] Davis mobilized a group of academic and federally employed geographers to create the Association of American Geographers in 1904. New members would have to demonstrate a record

---

to the best of his ability." Edwin A. Grosvenor to AGB (who was still in London), 31 July 1900, Box 99, Grosvenor Papers.

66 Grosvenor, "Story of the Geographic"; Grosvenor, *NGS and Its Magazine*; Bryan, *100 Years*; Pauly, "World and All." On the American News Company, see Peterson, *Magazines*.

67 John Hyde's last issue, at least according to the masthead, was October 1901.

68 T. W. Freeman, *A Hundred Years of Geography* (London: Gerald Duckworth & Co., 1961), 66. See also Pauly, "World and All." Among Davis's 11 articles for the magazine are "Geographic Methods in Geologic Investigation," *NGM* 1 (October 1888): 11–26; "The Rivers of Northern New Jersey, with Notes on the Classification of Rivers in General," *NGM* 2 (May 1890): 81–110; and "The Rational Element in Geography," *NGM* 10 (November 1899): 466–473.

69 Pauly, "Iconography of an Emerging World," 16; McGee to Favis, 29 May 1902, Harvard Geographical Institute Papers, Houghton Library, Harvard University.

of publication of scholarly material, and require approval by the Association's executive council. The formation of the Association of American Geographers as the new organizational home for "expert quality" geographers took some of the pressure off Grosvenor and allowed him to continue his slow, steady popularization of the *National Geographic Magazine*.[70]

Secure in the Board's recognition of Grosvenor's ability, Bell stepped down as president of the National Geographic Society in late 1903. "'Bert, you are competent to paddle your own canoe,'" Grosvenor recalled him as saying.[71] In 1905, the Society took over the payment of Grosvenor's salary, boosting it by a third, and attaching the next raise to the acquisition of an additional 15,000 members; Grosvenor was also unanimously elected to the board.[72] That same year, membership rose from 3,662 to 11,479, and Grosvenor hired his first assistant, John Oliver La Gorce, who would remain his right-hand man for more than 50 years.[73] Further securing his authority in 1907, Grosvenor acquired responsibility for "the general business of the Society," including all employees. The 1907 changes in the bylaws of the National Geographic Society essentially rendered the board "an honorary body composed primarily of Grosvenor's friends and relatives."[74] In 1920, by which time membership had reached 765,000,[75] Grosvenor became president of the National Geographic Society. He held this post, together with that of editor, until his retirement in 1954, when he became Chairman of the Board. While the staff grew enormously – in 1905 there were nine, and in 1938 there were 800 – Grosvenor was clearly in charge during all but the earliest days of his lengthy editorship. And although he would come to proclaim often that the magazine was by principle objective and nonpartisan, as the next chapter seeks to show, its politics became, by and large, his politics.

---

70 William Morris Davis, "Geography in the United States, I," *Science* NS 19 (January 1904), 126. Besides Davis, a few charter members of the AAG were also on the NGS Board of Managers – O.P. Austin, Henry Gannett, Grove Karl Gilbert (who resigned from the NGS in 1905), Angelo Heilprin, C. Hart Merriam, Rollin D. Salisbury (who resigned from the NGS in 1906), and W J McGee. After a brief episode as president of the NGS, McGee resigned from the Board of Managers in 1904. See Grosvenor, *NGS and Its Magazine*, 147; James, *All Possible Worlds*, 364; Hinsley, *Savages and Scientists*.

71 Grosvenor, *NGS and Its Magazine*, 41. Bell also says as much in AGB to GHG, 7 October 1905, Box 99, Grosvenor Papers.

72 GHG to AGB, 30 September 1905, and AGB to GHG, 7 October 1905, Box 99, Grosvenor Papers; Grosvenor, *NGS and Its Magazine*, 43.

73 Grosvenor, *NGS and Its Magazine*, 44; Bryan, *100 Years*, 95.

74 Pauly, "World and All," 526.

75 The "Report of the Director and Editor of the National Geographic Society for 1919" noted that the magazine's circulation equaled the combined circulations of the *Atlantic Monthly* (90,000), *Century* (50,000), *Harper's* (80,000), *Outlook* (100,000), *Review of Reviews* (200,000), *Scribner's* (90,000), and *World's Work* (140,000). Box 160, Grosvenor Papers.

Chapter 2

Picturing the World, Imagining the Nation

National Geographic's development needs to be seen within the context of a new American overseas imperialism and the emergence of a United States that was coming to see itself as a world leader in the arenas of economics, politics and civic culture. While actively participating in the rise of the U.S. as a global power, National Geographic, under the leadership of Gilbert H. Grosvenor, professed disinterest in politics, for the most part presenting a picture of global sunniness. This chapter examines National Geographic's relationship with the nation – both the United States government and the "imagined community" of Americans.

Through focusing on "the world and all that is in it," the *National Geographic Magazine* worked to forge among its American readers a particular *national* identity. Benedict Anderson, in *Imagined Communities*, provides a compelling argument for understanding nations as "imagined political communities," a communal act of imagination fomented by the invention of the printing press, the development of literatures in particular vernacular languages, and the spread of print capitalism.[1] The stories told through the media – the knowledge, the tropes, the ruminations, the exhortations – were and continue to be shared and reproduced throughout a national base that was constructed initially by geographical contiguity and held together, at least theoretically, by linguistic unity. Through everyday consumption of national media, an affluent Bostonian and a poor rural Texan, for example, can share a common cultural base that facilitates a connection regardless of space or situation and that enables them to see themselves as "Americans."[2]

At the time of the National Geographic Society's founding in 1888, the United States was developing an "American" national identity, a project that had been earlier undermined by a long period of furious sectionalism that culminated in the Civil War. After the war, the U.S. experienced a coalescence of national-shaping developments. Of particular importance was the completion of the transcontinental railroad system in 1869. The railroad enabled the late nineteenth-century growth of largely agricultural and ranching settlement in the plains and southwestern territories – at the expense of Native Americans and other previous inhabitants – that helped "fill in" the American frontier. The new ease of shipping goods allowed for nationwide

1 Anderson, *Imagined Communities*, 6.

2 Struggling Texan Lyndon B. Johnson and affluent Bostonian Leverett Saltonstall provide this example. See Afterword, p. 165.

expansion of markets beyond their extant local or regional scale. Local time, too, succumbed to the national scope of the railroad, with the creation of formal time zones across the continent in 1883.

The national magazine was also "born" in the late nineteenth century, helped along by the reduction of mailing costs for periodicals by the Postal Act of 1879, and the adoption of mass production methods in printing processes by the 1890s.[3] While the older "dignified" or "genteel" magazines, such as *Harper's Monthly* and *Scribner's*, catered to the northeastern establishment and its Victorian values and tastes, editors of newer magazines such as *McClure's*, *Cosmopolitan* and *Munsey's* cast a wider net, aiming for the "literate middle class" [white] in the midwest, west and south as well.[4] Circulation of the monthly magazines boomed from 18 million in 1890 to 64 million in 1905.[5]

Richard Ohmann credits late nineteenth-century magazines with creating American national mass culture, a significant component of which, he argues, was the formation of Americans as consumers. Regional companies were expanding into national markets and the magazines increasingly relied on national brand advertisers for income. Advertising allowed for lower subscription rates, which led to more people subscribing, which lured more companies to advertise in the magazines. Learning to see themselves as consumers was an integral part of what Ohmann describes as the development of a professional-managerial class in American culture and society, the prime audience for popular monthlies like *Munsey's*, *McClure's*, *Cosmopolitan* and the *Ladies' Home Journal*.[6]

Of course, the editorial part of the magazines – the stories and the illustrations – also had an impact on its readers. As Schneirov notes, "popular magazines helped to construct three new social identities for their readers – as consumers, as clients of professional expertise, and as educated, nationally oriented citizens."[7] For both editorial and advertising purposes, the magazines had to "imagine" their audiences as a national whole. The readers themselves became a community of sorts, sharing the information – and the cultural capital – gleaned from their acts of magazine-specific readership. By the early twentieth century, *National Geographic* was both a competitor of the popular monthlies and a distinctive outlier. As with the popular monthly magazines Bell and Grosvenor studied, the revamped *National Geographic Magazine* carried advertisements from respectable national brands, although the *Geographic* always segregated the ads, placing them either before or after the issue's content. The reminder to "Mention the Geographic – it identifies you" ran beneath advertisements in the *National Geographic Magazine* for over half a century.

A 1907 *Boston Herald* editorial cheered the *Geographic*'s role in shaping an imagined American community, fostering a national identity through imagining the

3 Peterson, *Magazines*; Ohmann, *Selling Culture*; Schneirov, *New World Order*.

4 Schneirov, *New World Order*, 97.

5 Ohmann, *Selling Culture*, 29.

6 Ibid.

7 Schneirov, *New World Order*, 124.

United States as a formidable and coherent entity, unique in its relation to the rest of the world. The newspaper noted that geography can be tedious, or, as the National Geographic Society has proved, it can be fascinating. Moreover, as practiced by the National Geographic Society, it can nurture and sustain patriotism. "Indeed, the modern incubation of hatching chickens by incubators instead of hens is simply nowhere compared with the system of hatching patriots of the stamp of William Tell by geological geography, as exemplified in the faith and works of the National Geographic Society of Washington, D.C." The editorial went on to say that patriotism necessitates a love of something and someones more concrete than vast numerical abstractions, and it credited the *National Geographic Magazine* for helping create national sentiment through its particularly *visual* portrayal of places. It used to be said that "The States were too big, too broadly dispersed, too divergent in interests, for anyone to be capable of loving their multitudinous populations as fellow-countrymen. All this, however, at any rate in the eyes of the National Geographic Society of Washington, is now rapidly being done away with. It is getting effected through a vivid appeal to the visual imagination which is enabling us all to see in the mind's eye the whole country at once and as a whole."[8]

Formulating *National Geographic*'s political vision

By the early twentieth century the dominant ideology of the Society and the magazine was, in keeping with the times, a form of Progressivism. Progressivism was primarily a political and social movement with a strong moral streak, a quest for improving society in large part through scientific expertise and efficient "good" government. While activists in the populist, socialist, labor, and anarchist movements called for different degrees and forms of radical social change, the middle- and upper-class Progressives called for "reform." Self-proclaimed Progressives subscribed to a fairly wide range of beliefs and objectives. The Progressive movement embraced "clean-government crusaders, conservationists, Anglo-Saxon supremacists, muckraking journalists, social-welfare workers, efficiency experts, middle-class professionals, and advocates of business regulation."[9] In general, however, Progressivism was a "Yankee-Protestant" middle-class and patrician movement, the political manifestation of the professionals' attempts to clean up a society made messy by rapid urbanization

8 *Boston Herald*, 23 November 1907, reprinted under the heading "Imagination and Geography" in *NGM* 18 (December 1907), 825. Grosvenor highlighted the magazine's depiction of the United States in a promotional "give your friends a subscription" letter, late 1908–early 1909, stating that not only does the magazine present "vivid descriptions of strange countries and stranger people [and] actual pictures of their daily life, the fireside, the work and the play," but also "the record of our country's onward giant stride, for instance, the progress of the great reclamation projects of the Government in the West and its plan for Forest preservation all over the United States." Box 160, Grosvenor Papers.

9 Emily S. Rosenberg, *Spreading the American Dream: American Economic and Cultural Expansion 1890–1945* (New York: Hill & Wang, 1982), 41–42.

and industrialization, heavy immigration from non-English-speaking European countries, and the abandonment of what many saw as democratic business ethics by cartels and corporations.[10]

Embroiled in American politics, with its overlapping layers of local, state and federal government, Progressivism was attuned to issues of scale. It began largely as a local and state movement, calling for the reform of municipal governments in particular. Through the introduction of efficiency and expertise, such as the civil service system, with regulated exams tied to each position, instead of the prevailing government-job-as-reward practice, Progressives fought against government corruption. This also generally meant attacking the increasing electoral dominance of immigrant populations who supported the political machines. Aligned with Progressivism was the "City Beautiful" movement, which called for slum clearance and the construction of grand public buildings and boulevards, another effort of "middle-class reformers ... to impose social control and moral order."[11] The U.S. acquisitions of 1898, with their attendant questions of governance, gave Progressives a whole new scale of operation, and McKinley's assassination in 1901 at the Buffalo World's Fair – by an anarchist, no less – placed an outspoken Republican Progressive in the White House: Theodore Roosevelt.

Roosevelt, contributor of seven articles to the *National Geographic Magazine*, and the Republican founder of the short-lived Progressive Party, formed for the 1912 presidential elections, is remembered as a reformer in terms of his conservationism, anti-trust action and regulatory legislation. But the man for whom the teddy-bear was named was also an Anglo-Saxonist warmonger and expansionist.[12] Indeed, although many progressives today find imperialism incompatible with their beliefs, many early twentieth-century Progressives supported U.S. imperialism. Progressivism shared with imperialist ideology the belief that the world would be improved if the educated, energetic professionals of the most advanced culture took over. "To a considerable degree, American imperialism of the early twentieth century, especially during the Republican ascendancy of Roosevelt and Taft, was progressivism writ large."[13]

---

10 See Richard Hofstadter, *The Age of Reform: From Bryan to F.D.R.* (New York: Vintage, 1955), 9, who uses the term "Yankee-Protestant"; Arthur A. Ekirch, Jr., *Progressivism in America: A Study of the Era from Theodore Roosevelt to Woodrow Wilson* (New York: New Viewpoints, 1974); Robert H. Wiebe, *The Search for Order: 1877–1920* (New York: Hill and Wang, 1967); Rosenberg, *Spreading the American Dream*; and Brands, *Bound to Empire*.

11 Richard E. Fogelsong, *Planning the Capitalist City: The Colonial Era to the 1920s* (Princeton: Princeton University Press, 1986), 137.

12 Brands, *Bound to Empire*, 14–16, devotes a few amusing paragraphs to demonstrating how Roosevelt "valued war for its own sake," citing several documented inflammatory statements by and about Roosevelt, as well as an unattributed quote from a college friend that Roosevelt "wants to be killing something all the time... . He would like above all things to go to war with some one."

13 Brands, *Bound to Empire*, 61. Ekirch, *Progressivism in America*, makes much the same point.

The ideological zone in which Progressivism aligned itself with imperialism lasted long beyond the "Progressive Era," commonly framed by the first 16 years of the twentieth century.[14] It can be seen as part of a largely economically motivated ideology that historian Emily Rosenberg calls liberal-developmentalism, which lasted from the late nineteenth to the mid-twentieth century. The prevailing political-economic ideology in the U.S., liberal-developmentalism couched economic imperialism in terms of altruism, evolution, and world progress.[15] Rosenberg identifies five tenets of liberal-developmentalism:

> 1) belief that other nations could and should duplicate America's own developmental experience; 2) faith in private free enterprise; 3) support for free or open access for trade and investment; 4) support for free flow of information and culture; 5) growing acceptance of governmental activity to protect private enterprise and to stimulate and regulate American economic and cultural exchange.[16]

Within the rhetoric of liberal-developmentalism, the United States stood as a beacon of the political ideals enshrined in the country's laws and myths of origin. These ideals were liberty and freedom and the duty of the U.S. was to carry its enlightened system to oppressed countries everywhere. The "freedom" to be spread, of course, was that of the "free market." Rosenberg charts the growth of this ideology from the last decade of the nineteenth century, when to many business leaders and politicians, expansion seemed an obvious solution to the looming problems of overproduction and domestic unrest. The success of the Spanish-American War of 1898 secured liberal-developmentalism as the dominant framework for U. S. foreign relations. "Economic need, Anglo-Saxon mission, and the progressive impulse joined together nicely to justify a more active role for government in promoting foreign expansion."[17] The 1912 Progressive Party platform, for example, called for the federal government to "co-operate with manufacturers and producers in extending … foreign commerce."[18]

The *National Geographic Magazine* actively promoted U.S. involvement in foreign territories. At the beginning of the twentieth century, the active membership and leadership of the National Geographic Society was still largely held by employees of the federal government. As Grosvenor ascended into the editorship, he continued to tap the resources of information he had in the persons of the associate editors,

---

14 For estimations of the duration of the Progressive Era, see Dubofsky et al, *The United States in the Twentieth Century* (Englewood Cliffs, NJ: Prentice-Hall, 1978); Hofstadter, *Age of Reform*; Ekirch, *Progressivism in America*.

15 Rosenberg, *Spreading the American Dream*.

16 Ibid., 7.

17 Ibid., 42.

18 Quoted in Burton I. Kaufman, *Efficiency and Expansion: Foreign Trade Organization in the Wilson Administration, 1913–1921* (Westport, CT: Greenwood Press, 1974), 67.

many of whom headed federal departments.[19] The National Geographic Society's school bulletin, initiated in 1919, was "the only non-government publication ever sent out regularly under the government frank."[20] And although by the 1930s, the magazine was offering far fewer reports from U.S. bureaus, the Geographic prided itself on the tradition of having the current U.S. president, regardless of political party, bestow the Society's Hubbard medal on its honorees.[21] The Geographic had a particularly warm relationship with Theodore Roosevelt, including covering his hunting trip in Africa during his short retirement from politics. Grosvenor was also friendly with Calvin Coolidge – they knew each other from Amherst College, where Grosvenor had considered Coolidge "a worthy opponent in handball"[22] – and was second cousins with William Howard Taft. Taft, the U.S. President and then Supreme Court Justice, was governor of the Philippines from 1901 to 1904. In 1904, Taft moved back to Washington and into the post of U.S. Secretary of War, the head of the federal department ultimately responsible for the Philippines. Taft contributed 16 pieces for *National Geographic*, three on the Philippines.

National Geographic, as was the case of many of the most popular magazines of the early twentieth century,[23] was firmly ensconced in the particular cluster of ideas that corresponded to a "progressive conservative" ideology.[24] This mélange of nationalism, adventure capitalism, Anglo-Saxonism and general racism, social evolutionism, moral righteousness, missionary Protestantism, veneration of efficiency and productivity, and what Donna Haraway calls "teddy bear patriarchy,"[25] provided the ideological setting for *National Geographic*'s interpretation of the world to its readers.

According to Rosenberg, a linchpin of liberal-developmentalism was "faith in the ability of Americans to perfect and apply laws of progressive betterment and to

19 "I would recommend you to cultivate your Associate Editors in the Departments and through them get the publications of the Departments," Bell wrote to Grosvenor in 1901. "There is an immense amount of matter published relating to subjects suitable for the magazine." AGB to GHG, 31 January 1901, Box 267, Bell Papers.

20 "For two years the U.S. Bureau of Education mailed them out but the demand for them became so great that the Bureau had to relinquish that task." From notes for Gilbert H. Grosvenor, "Some Recent Work of the National Geographic Society" 21st annual meeting of the American Association of American Geographers, Washington, D.C., 30 December 1924 – 1 January 1925, Box 159, Bell Papers.

21 Ross, "Geography, Inc.," 26.

22 Grosvenor, "Story of the Geographic," 50.

23 Ekirch, *Progressivism in America*, 187, cites *The Independent, Outlook, Century, Harpers,* and the *North American Review* as examples, noting their frequent articles and editorials "in support of imperialism, not only in terms of American economic interests, but as the democratic duty and world responsibility of the United States."

24 According to Judith Icke Anderson, *William Howard Taft: An Intimate History* (New York: W. W. Norton, 1981), 138, Roosevelt referred to himself as a "progressive conservative."

25 Haraway, *Primate Visions*.

uplift those lower on the evolutionary scale."[26] Permutations of evolutionary ideas had been used to justify imperialism before, most notably in the theory of social Darwinism. As Richard Hofstadter noted, "Darwin was talking about pigeons, but the imperialists saw no reason why his theories should not apply to men, and the whole spirit of the naturalistic world-view seemed to call for a vigorous and unrelenting thoroughness in the application of biological concepts."[27] Spencer's social application of "survival of the fittest," for example, provided a naturalized justification for colonial conquest; conquering peoples were only following the laws of nature, and what was natural was both inevitable and appropriate. Social evolution, therefore, allowed for framing imperialism and colonialism as natural processes, and in so doing, is an important "strategy of innocence" for anti-conquest narratives. Social evolution fed into U.S. liberal-developmentalism, with its belief in the United States as a "vanguard of world progress,"[28] and allowed American writers to "secure their innocence" through naturalizing the United States' position of power in the world.

Concepts of racial hierarchy were integral to ideas of social evolution. What was implicitly assumed was that some races were hereditarily weaker than others and so merited conquest, or that some races were more culturally advanced than others and so were responsible for "helping" the others to progress, taking on the "white man's burden."[29] Ideas of racial superiority and inferiority had long predated evolution, of course; slavery and genocide did not need Darwin or Lamarck. The concept of race was used to mark difference, and in the U.S. around the turn of the century, that difference could have biological, cultural and geographical markers, and usually, but not always, all three.

Both hereditary and culturally-based concepts of racist thought played into anti-immigrant arguments as well as into positions for and against U.S. overseas expansion. Anti-imperialist arguments were used to attack American hypocrisy regarding republican ideals, and to expose U.S. economic greed. But they were also used by those who feared "the admission of alien, inferior, and mongrel races to our nationality."[30] The *National Geographic Magazine* tended to support the "uplift" concept of culturally-based racist hierarchy, which emphasized the ability of the inferior to *learn* to achieve higher levels of civilization. But underlying the cultural aspect was a clear hereditary concept of race, where the "uplifting" process could take a hundred generations, thereby rendering little threat to the status quo.[31] Under the leadership of Gilbert Grosvenor, the magazine made few alterations to

26 Rosenberg, *Spreading the American Dream*, 9.

27 Hofstadter, *Social Darwinism,* 171.

28 Rosenberg, *Spreading the American Dream*, 9.

29 See Hofstadter, *Social Darwinism*; Stephen Jay Gould, *The Mismeasure of Man* (New York: W.W. Norton and Company, 1981).

30 The quotation is from E. L. Godkin, editor of the *Nation*, who apparently was one of those all-argument embracing anti-imperialists. Quoted in Brands, *Bound to Empire*, 27.

31 Theodore Roosevelt was able to balance his beliefs in racial difference with his belief in a unitary American nationality through this gradual change formula. See Richard Slotkin,

its generally racist ideology of social evolution, presenting its readers a visually-enhanced narrative of "World Progress" in which a strong and benevolent United States showed the way.[32]

Anglo-Saxonism only fine-tuned definitions of the racial-social hierarchy so that the "British race" (Scots and Welsh generally included, Irish generally excluded) was at the top of the hierarchy. American Anglo-Saxonism trumpeted the virtues and the victories of British achievement, particularly the settlement of North America, where the admixture of good Anglo-Saxon stock and blessed geography led to the rise of the United States. Gilbert Grosvenor, a member of the English-Speaking Union, proudly asserted that "every drop of blood" of his was from English ancestors.[33] He remarked once that "The British race (by that I mean the men living either in the British Isles or of British descent in America) has shown the greatest inventive genius any people have ever had."[34]

Protestant ministries, especially those with substantial missionary interests, were generally supportive of, and often overjoyed about, U.S. overseas expansion. In referring to the religious rationale for expansionism as "the imperialism of righteousness," Julius W. Pratt noted that, "It would seem that the only Christian denominations that were genuinely and thoroughly opposed to [the Spanish-American] war were the Friends and Unitarians."[35] The fact that 90 percent of Filipinos were Catholics did not deter the Protestant faithful from calling on their brethren to "Christianize" the Philippines, while American Catholics were generally eager to show their loyalty to the United States and its political system. As Rosenberg explains, "Most Americans believed that Protestant Christianity was a spiritual precondition for modernization."[36] This belief was exemplified by William H. Taft's report, abstracted in the *National Geographic Magazine*, on his "Ten Years in the Philippines." The Catholic friars, said Taft, left the Filipinos in a state of Christian tutelage, "ripe to receive modern western conceptions as they should be educated to

---

Lost Battalions: The Great War and the Crisis of American Nationality (New York: Henry Holt, 2005), 22.

32 Lutz and Collins, *Reading National Geographic*, find much of the same in their examination of the magazine after 1950.

33 Poole, *Explorers House*, 39, quoting a 1962 interview.

34 GHG to AGB, 14 May 1917, Box 100, Grosvenor Papers. Grosvenor's mother-in-law, apparently, favored "Australasia" as the proving ground "for folk of our own race." In a talk she prepared and Elsie Grosvenor delivered to a 1912 Washington Club Tuesday Morning Current Events Meeting, Mabel Bell exhorted her audience to "Think of all our forefathers had to contend against in opposing Spaniards, French and Indians, and of our own struggles against the on-coming tide of alien immigration," and compare that with the isolated and "almost uninhabited territory" that the British and their descendants came to and maintained in Australia and New Zealand (27 February 1912, Box 61, Bell Papers).

35 Pratt, *Expansionists of 1898*, 288, 293. See also Edward McNall Burns, *The American Idea of Mission: Concepts of National Purpose and Destiny* (New Brunswick: Rutgers University Press, 1957).

36 Rosenberg, *Spreading the American Dream*, 8.

understand them. This is the reason why I believe that the Christian Filipino people are capable by training and experience of becoming a self-governing people."[37]

Anglo-Saxonists on the hereditary end of the racism spectrum were often supporters of the eugenics movement, although eugenicists were not necessarily Anglo-Saxonists. Eugenics held a place among the scientific reforms of the Progressive Era. Supporters believed that the science of heredity, and social policy informed by such science, would lead to a better society by eliminating inferior strains of humanity (a category open to interpretation, even among eugenicists) and encouraging the procreation of superior strains of humans. The *National Geographic Magazine* ran three articles on eugenics, two of which were authored by Alexander Graham Bell.[38] Bell came to eugenics through his research on hereditary deafness, and spent years breeding multiple-nippled sheep.[39] He became a prominent member of the American Breeding Association's Committee on Eugenics, established in 1906 "to investigate and report on heredity in the human race [and] to emphasize the value of superior blood and the menace to society of inferior blood."[40]

Bell was more interested in the science of heredity than in racial ideology, but other scientists involved with the Committee on Eugenics were social activists. Harvard climatologist Robert De Courcy Ward was a leading advocate of immigrant restriction, and *National Geographic* published his article "Our Immigration Laws from the Viewpoint of Natural Eugenics" in 1912. We have enough "degenerates" already here, he argued. After all, we are careful about importing cattle, why should we not be careful about humans? "Natural eugenics, for us, means the prevention of the breeding of the unfit native, as well as the prevention of the immigration and of the breeding after admission of the unfit alien."[41] Another eugenics activist who contributed to the magazine was Madison Grant, trustee of the American Museum of Natural History, secretary of the New York Zoological Society, and co-founder of the California Save-the-Redwoods League. His *National Geographic* article focused only on rallying support for the redwoods, but Grant's interest in preserving

37 William H. Taft, "Ten Years in the Philippines," *NGM* 19 (February 1908), 142.

38 The two articles are "A Few Thoughts Concerning Eugenics," *NGM* 19 (February 1908): 119–123; and "Who Shall Inherit Long Life: On the Existence of a Natural Process At Work Among Human Beings Tending to Improve the Vigor and Vitality of Succeeding Generations," *NGM* 35 (June 1919): 505–514.

39 Dorothy Harley Eber, *Genius at Work*, (New York: Viking, 1982); Mark H. Haller, *Eugenics: Hereditarian Attitudes in American Thought* (New Brunswick: Rutgers University Press, 1984). Bell, whose life work was dedicated to what would now be called "mainstreaming" the deaf, by teaching them speech, concluded from his study of familial deafness that it would be unwise for two congenitally deaf people, or two people from families in which more than one person was deaf, to marry.

40 Haller, *Eugenics*, 62. Poole, *Explorers House*, 63, says that Bell dropped his involvement with the committee when he realized its thrust was more racist than scientific.

41 Robert De C. Ward, "Our Immigration Laws from the Viewpoint of Natural Eugenics," *NGM* 23 (January 1912), 39. Ward exulted when the U.S. Congress passed a law excluding immigrants from much of Asia (Haller, *Eugenics*, 154).

the glories of nature extended to his belief that sterilization, selective breeding and apartheid techniques should be used to keep the strongest races undiluted.[42] As Donna Haraway notes, eugenics and conservation "all seemed the same sort of work" within a certain strain of Progressivism.[43] It was a Progressive vision of an ideal state "supported by a ruling group of savants and luminaries who were in full possession of scientific facts, sufficient for them to make policy decisions" regarding those deemed racial others as well as "perverts, bastards, cripples, and feeble-minded individuals living in the new intellectual and industrial state."[44] As Ward put it, the goal of eugenics was "conservation of the American race."[45]

The National Geographic Society showed more interest in conservation than in eugenics, generally following the federal policy line of resource management.[46] Grosvenor was quite proud of the Society's role in preserving a swath of redwoods and sequoias in California, although his motives seem to have stemmed as much from advancing the Society's reputation as from any ecological concern. Explaining to Bell why he thought the Society should spend a large chunk of its research fund to help the U.S. government purchase and protect the giant-tree covered land near Sequoia National Park, Grosvenor argued that the investment would ultimately bring great benefit to the NGS. Invoking the Society's close ties to the federal government, Grosvenor reasoned that:

> The National Geographic Society is generally regarded as semi-official, and our joining with the Government in rescuing the trees would be a dignified and patriotic proceeding. Furthermore, we have more members in California, in proportion to its population, than in any other state, and I think the appropriation, if made, would enable us to get several thousand more members in that state. If this $20,000 were spent on two or three expeditions the results would be more or less ephemeral, whereas if invested in the big trees it would be an investment for generations. The Government would put up a tablet in the Park commemorating the action of the National Geographic Society. The Society would also receive a great deal of publicity for its action, which is always helpful.[47]

Rather than purchasing the land outright, and holding ownership over it, the National Geographic Society instead chose beneficence and patriotism and so further identified the Society with the federal government as "good government."

The conservation movement was in many ways a Progressive response to fears of unchecked capitalism, the "closing" of the American frontier, and unscientific

---

42 Ivan Hannaford, *Race: The History of an Idea in the West* (Washington, D.C. Woodrow Wilson Center Press, 1996). See also Haller, *Eugenics*. Hannaford names Grant as a trustee of the National Geographic Society, but Grant's name does not appear on a list of all the members of the Society's Board of Managers, later the Board of Trustees (Grosvenor, *NGS and Its Magazine*, 147).

43 Haraway, *Primate Visions*, 57.

44 Hannaford, *Race*, 358.

45 Ward, "Immigration Laws," 41.

46 Pauly, "Iconography of an Emerging World," 17–18.

47 GHG to AGB, 26 October 1916, Box 100, Grosvenor Papers.

management of resources. "Conservation of our natural resources is only preliminary to the larger question of national efficiency, the patriotic duty of insuring the safety and continuance of the nation," Theodore Roosevelt told his 1908 Conference of Governors. This concern about resources in the national interest "was the way the United States, under the policies of imperialism, was turning increasingly to the exploitation of natural resources and native peoples of the hitherto less developed areas of the world."[48] Indeed, in the early twentieth century the use, management, and geography of "resources" had become a prominent issue. Geography textbooks, for example, changed in focus from physical geography to commercial geography after 1898, emphasizing the natural resources of particular countries and regions. In turn, a country's or a people's level of civilization was considered tied to the degree to which they made efficient use of available resources, a measure determined solely by outsiders who wanted those resources. If resources were not used efficiently, then it was incumbent upon more advanced peoples to come in and avail themselves of these valuable elements.[49]

The United States created a most important resource for itself when it beat out European interests in the region to construct what became the Panama Canal. The land on either side of the planned canal, the Canal Zone, became formal U.S. territory in 1903, and the canal opened for business in 1914. The country of Panama itself, a breakaway province of Colombia whose revolution was fostered by the United States, was immediately ushered into the United States' "sphere of influence." A 1922 *National Geographic Magazine* article on Panama shows that the magazine was still a forum for questions regarding how well the white race would function in the tropics, a point that had long concerned European colonizers. The United States, with the advantage of its late entry into the imperial arena and advanced scientific techniques, appeared to be faring better than its European counterparts in the tropics. David Fairchild, a U.S. Department of Agriculture botanist and Gilbert Grosvenor's brother-in-law, reported:

> [T]his is the first time in history that a northern race has comprehended, and shown that it comprehended, the gigantic scale upon which it will be necessary to operate if the white races ever conquer the tropics.
>
> Much has been said about the inability of the white race to live there. Perhaps it cannot live there as the brown and black races do; but for all of that it can and will accomplish great changes; and the development of the Panama Canal Zone, with its sanitation, transportation facilities, its admirable hotels, and its stirring intellectual life, stands as a brilliant example of what the future may bring in the development of the gigantic resources of the topics.
>
> It is from this standpoint that one should view the accomplishments of our country and urge it to go on with the research work which it has begun, and make here, in this

48 Ekirch, *Progressivism in America*, 149–150.

49 Schulten, *Geographical Imagination in America*.

> frontier post, the discoveries without which the scientific conquest of the tropics will be impossible.[50]

In his article, Fairchild makes clear that the expert level of science at work in the Panama jungle did not interfere with a boy's (or a man's) primal joy at encountering wild nature. "'Me for the tropics!' was the boy's exclamation," when his father brought him to Panama. Fairchild had decided to travel with his son, Alexander Graham Bell Fairchild, to the Panama Canal Zone jungle because, as he asserts in the opening sentence of the article, "The more I thought about it, the more it seemed to me important that my boy should, before his habits of thought and life had become conventionalized, feel the grip of one of the most tremendous of all experiences, that of being all alone in a tropical jungle."[51]

The "boy in wild nature" theme has long been prominent in English and American literature, part of a larger genre that could be referred to as "stranded white Christian male in an unknown landscape," exemplified by Daniel Defoe's *Robinson Crusoe*. Richard Phillips, in his study of imperialist adventure fiction, including *Robinson Crusoe*, shows how "for the most part the eclectic geographical literature of adventure is restricted to *male* encounters – real and imaginative – with nature and the unknown."[52] Throughout the Victorian period and well into the twentieth century, tales of adventure and exploration were considered "boy's stories," even though girls comprised a significant portion of the readership. While Fairchild alludes to the "timeless" quality of a boy discovering his energetic and masculine self through confronting wild nature, the context is clearly larger, that of American scientific imperialism. The boy's future role – he is to be a scientific American – is mapped into the landscape as Fairchild depicts it.[53] Indeed, *National Geographic* fostered an ideal of American masculinity as one of both physical endurance and scientific/technical endeavor, as exemplified by explorers such as Robert Peary, whose polar expedition was the National Geographic Society's first sponsored expedition.[54]

Donna Haraway identifies a slightly earlier version of American imperialist masculinity that she terms "teddy bear patriarchy," named for its great exemplar, Theodore Roosevelt. In her analysis of the American Museum of Natural History's Theodore Roosevelt memorial atrium, and the museum's activities and aspirations

---

50 David Fairchild, "The Jungles of Panama," *NGM* 41 (February 1922), 144. Fairchild, married to Alexander Graham Bell's daughter Marian (Daisy), is described in his byline here as "Agricultural explorer in charge of foreign seed and plant introduction, Department of Agriculture."

51 Ibid. 131.

52 Richard Phillips, *Mapping Men And Empire: A Geography of Adventure* (London: Routledge, 1997), 45.

53 As it turns out, Graham Fairchild felt the grip so strongly that he became an entomologist and spent over 30 years living and doing research in Panama. [http://en.wikipedia.org/wiki/Graham_Fairchild, accessed 22 August 2006; "Alexander Graham Bell Fairchild, 87, Dies," *The New York Times*, 17 February 1994.]

54 Bloom, *Gender on Ice*.

in the first decades of the twentieth century, Haraway calls the museum "a space that sacralizes democracy, Protestant Christianity, adventure, science, and commerce."[55] Though different forms of media, the museum and the *National Geographic Magazine* carried the same message. For Roosevelt, and others like him, "the great outdoors" was a necessary proving ground for males made "soft as girls" by city life,[56] or turned decadent by the trappings of wealth and extravagance. Roosevelt called for "principles of virile honesty and robust common sense … in our civic life,"[57] attaching masculine virtues associated with outdoorsmanship to his concept of nation.

"Teddy bear patriarchy" was a kind of macho paternalism. In its official imperialism, the United States adopted the stance of a strong father: forceful, protective, and eager to raise his "children" to adulthood, but neither wanting or expecting to ever lose authority. In an era when "primitive" peoples were often considered "children of nature," as they were in a number of *National Geographic* articles in the early twentieth century,[58] non-Western people living under Western imperial or colonial domination were considered political children. In the Progressive terms of governmental "uplift," children needed the ever-watchful guidance of their more civilized parent. This attitude is apparent in articles such as Taft's "Ten Years in the Philippines." For example, immediately following his statement that he believes the Filipinos have what it takes to become a self-governing people, Taft argues that the United States' continued control of the Philippines is necessary because "for the present day they are ignorant and in the condition of children."[59] The metaphor of the parent-child relationship was favored by Progressive imperialists, for whom a master-subject relationship threatened untouchable democratic ideals. Many U.S. citizens found the position of the United States as an imperial nation to be a contradiction of American ideals. If the United States was anti-colonialist and pro-liberty, how did one explain the events and aftermath of 1898? More than two decades later, in a civics course for newly-franchised women, University of Chicago professor J. W. Garner struggled to explain the United States' various relations with its dependencies. Referring in one of his lectures to the dependencies as "Uncle Sam's step-children," Garner noted that while the Filipinos were not U.S. citizens, "I dislike to call them subjects, because the term is so offensive to people in a Republican country. We associate the idea of a 'subject' with a person who lives under a monarchal regime, and yet I see no reason why we should not be frank with ourselves and say these people are subjects rather than citizens."[60] Still, for many

---

55 Haraway, *Primate Visions*, 27.

56 Quoted in Phillips, *Mapping Men and Empire*, 55, from an 1891 Canadian adventure book by E. Roper (F.R.G.S.), *By Track and Trail*.

57 Quoted in Ekirch, *Progressivism in America*, 131.

58 See Rothenberg, "Voyeurs of Imperialism."

59 Taft, "Ten Years," 142.

60 J.W. Garner, "The Carrie Chapman Catt Citizen Course: Some More About Our Dependencies; How We Came By Them and How They Are Governed," *The Woman Citizen*

Progressives, the rhetoric of patriarchal altruism provided a comforting alternative to frankness.

The continuing importance of the Philippines

The Philippines have the distinction of figuring prominently in three hallmark episodes in the history of the *Geographic*. The first, of course, was the initial U.S. acquisition of the Philippines. In June 1898, the *Geographic* published the "Philippine Number," which featured, among other photographs, a few of barely dressed native peoples. The second episode, in 1903, led the issue of bare-breasted women to be settled as a matter of policy. Grosvenor consulted with Bell regarding whether or not to publish a photograph of two bare-breasted Filipina women harvesting rice in the Calamianes Islands of the Philippines. Bell and Grosvenor concluded that "prudery should not influence the decision" since the pictures were "a true reflection of the customs of the times in those islands."[61] That it was photographs from the Philippines and not some other location that prompted an established policy regarding nudity was most likely coincidental. Still, as it happened, *National Geographic*'s policy favoring the display of bare-breasted women was intimately related to the U.S. possession of overseas territory.

The third Philippines-related watershed in *National Geographic*'s early history involved Taft and a hefty set of photographic plates from the first census of the Philippines. By 1905, Grosvenor had a well-established reciprocal relationship with several government departments that regarded the *Geographic* a useful venue for their work. For Grosvenor, "It was like striking gold in my own backyard when I found that Government agencies would lend me plates for illustrations in The Magazine."[62] Free use of pre-made photographic plates not only saved the Geographic money in terms of acquiring photographs, but saved the magazine the cost of turning the photographs into printable half-tones. The Philippine census had been initiated in February 1903, and two years later, the 3,500-page, illustrated four-volume report was ready for publication. Taft told Grosvenor about the upcoming report, adding that "the National Geographic Society could help the Government and the people of the Philippines" if Grosvenor published an article about the document. "Mr. Taft had distinctly said 'photographs,' and that word had become as musical to my ear

(28 August 1920), 342. The lecture printed in the August 21 issue was called "The United States and its Dependencies: Uncle Sam's Step-Children."

61 Grosvenor, *NGS and Its Magazine*, 27, 39. The photograph in question had been taken by Dean Worcester, a University of Michigan ornithologist who conducted collecting expeditions in the Philippines and was appointed a member of the Philippines Commission (headed by Taft) in 1900. See Brands, *Bound to Empire*. Worcester's 1912 article "Head Hunters of the Luzon," featured photographs taken by a U.S. Army sergeant on his staff, Charles Martin. Martin's photographs so impressed Grosvenor that he hired him in 1915 to head the *Geographic*'s new photographic laboratory. Bryan, *100 Years*, 128.

62 Grosvenor, *NGS and Its Magazine*, 37.

as the jingle of a cash register to a businessman," Grosvenor recalled.[63] He selected 32 plates in all, across a range of layout patterns and subject matter: the supervising Filipino elite, well dressed in Western styles and formally posed, rural villagers in a variety of ethnographic poses, boats, dwellings and examples of agriculture and industry.[64] The issue of the *National Geographic Magazine* in which "A Revelation of the Filipinos" appeared boosted the number of member-subscribers so significantly that Grosvenor ordered a second pressing.[65]

The Census itself was a remarkable act of colonial administration, "both a confirmation and a means of consolidating the 'pacification' of the archipelago,"[66] in the name of (social) science and order. In addition to American officials, more than 7,500 Filipinos, mostly provincial and municipal elites, were employed as supervisors and enumerators. This arrangement was heralded as "the first attempt on the part of any tropical people in modern times to make an enumeration of themselves."[67] All participants were required to take an oath of allegiance to the U.S. government.[68]

The census and its widely heralded use of Filipino administrators served as a public relations coup for the U.S. government, which needed it. In the wake of the United States' "liberation" of the Philippines from Spanish dominion, in expectation that a free republic would be recognized by the U.S., Filipino nationalists led by Emilio Aguinaldo had established a republican Philippine government. But the new Philippine Republic was not recognized by either the United States or by Spain, which, in the post-war treaty, sold the Philippines to the United States for $20 million. To stamp out the insurgent nationalists, the U.S. waged war in the Philippines from 1899 to 1902. After decimating two successive insurgent capitals, the U.S. Army drove the Filipino nationalists underground. In May 1900, Filipino guerillas attacked U.S. soldiers at Donsol; in response, the U.S. Army destroyed nineteen towns and 800 villages. Stories of brutal campaigns and torture filtered back to the United States, as did photographs of the carnage.[69] One of the American Army officers charged with heinous behavior towards Filipinos dismissed the accusations, telling a U.S. reporter that the Filipinos are "as a rule, an illiterate, semi-savage people, who

63 Ibid., 43.

64 Gilbert H. Grosvenor, "A Revelation of the Filipinos," *NGM* 16 (April 1905): 141–192. The article was subtitled "The Surprising and Exceedingly Gratifying Condition of Their Education, Intelligence, and Ability Revealed by the First Census of the Philippine Islands, and the Unexpected Magnitude of Their Resources and Possibility for Development."

65 Grosvenor, *The NGS and Its Magazine*, 44.

66 Vicente L. Rafael, "White Love: Surveillance and Nationalist Resistance in the U.S. Colonization of the Philippines," in *Cultures of United States Imperialism*, eds. Amy Kaplan and Donald E. Pease (Durham, NC: Duke University Press, 1993), 188.

67 Grosvenor "Filipinos," 140.

68 Rafael, "White Love."

69 See *Savage Acts: Wars, Fairs and Empire*, a film by the American Social History Project (Pennee Bender, Joshua Brown, and Andrea Ades Vasquez, directors), Center for Media and Learning, City of the University of New York, 1995.

are waging war, not against tyranny, but against Anglo-Saxon order and decency."[70] In May of 1902, the last of the nationalist revolutionaries surrendered, and on July 4, 1902, President Theodore Roosevelt declared the end of the Philippine War, although U.S. troops would continue to fight Filipino insurgents for the next ten years.[71] The Philippine Census was commissioned shortly thereafter, part of the U.S. project of "benevolent assimilation predicated on the simultaneous deployment and disavowal of violence."[72]

The Census presented, in text and in picture, a calm and peaceful Philippines. The Philippines were shown to be rich in resources and with enough native talent to assist the U.S. in exploiting the islands' resources to the fullest. The Census demonstrated that the Philippines were functioning as a "Progressive laboratory," with striking advancements in education, sanitation, public health, agriculture, and politics, all courtesy of the United States.[73] The subtext heralded a peaceful and righteous America that had triumphed in its first concerted effort at overseas imperialism.[74] The Census report was enormous, however, and expensive to produce. "[O]nly 4,000 copies, which were exhausted even before publication" were made. By publishing a summary of the Census report, complete with "a large number of the exceedingly beautiful pictures with which the report is illustrated," *National Geographic* served the interests of the U.S. government. Reproduction of the official census photographs in the *National Geographic Magazine* would come to supersede the newspaper pictures of carnage and destruction in Philippine towns and villages, and the story of U.S. "altruism" could be told once again. The popularity of that issue of the magazine suggests that it was a story that many American readers, or at least the Geographic's core membership – by this time largely teachers and upper- and middle-class professionals – wanted to hear and see.

Gilbert H. Grosvenor: The editor and his politics

A comfortable member of the East Coast Protestant elite and for most of his life a registered Republican (the exception came during Woodrow Wilson's presidency), Grosvenor in the early twentieth century was, in the manner of Theodore Roosevelt,

70 Quoted in Brands, *Bound to Empire*, 58.

71 *Savage Acts*, 1995.

72 Rafael, "White Love," 186.

73 Grosvenor, "Filipinos." Brands, *Bound to Empire*, 61, notes that "the Philippines, the primary proving ground for imperialism, also became a laboratory for progressivism."

74 In the *National Geographic Magazine*, the census report was followed by an article which stressed the importance of economic/commercial geography, arguing that commerce was the route by which the world would find peace, and proclaiming that "the time has come when the forces of commerce are being summoned to the suppression of the brute element of man, from which war and warfare are generated." Edward Atkinson, "Some Lessons in Geography," *NGM* 16 (April 1905), 197.

a "progressive conservative."[75] On the "progressive" side, Grosvenor framed his life's work as that of democratizing geographic education for a more informed citizenry. In his personal political life, he supported his wife's suffragist efforts and sent his daughters to a Montessori school.[76] But for the most part, Grosvenor was conservative. Theodore was the Roosevelt most favored by Grosvenor and his staff. By the time of Franklin D. Roosevelt's fourth election in 1944, the Republican bias in National Geographic headquarters was a source of frustration for the minority of New Dealers on the staff. "Much Dewey opposition caused a few of us to finally desert our roles of self-imposed silence, and set up FDR headquarters in our own small corner … which we did with an appropriate picture from Coronet and a resurrected 1940 campaign button, mascot attached," reported one Democratic National Geographic employee. "As a result, we discovered other friends. Mr. H. welcomed his immediate staff to the donkey-fold and vice-versa. Of course, we were *much* in the minority, but a most happy minority after the election."[77]

Grosvenor's beliefs made themselves felt at National Geographic headquarters. He did not drink or smoke, and made it company policy not to accept tobacco or alcohol advertising. Grosvenor also backed a policy designed by La Gorce to deny membership to African-Americans, at least in Washington D.C., a practice made feasible by the requirement for new members to be nominated. Grosvenor didn't want blacks to use the library, vote at annual membership meetings, or appear at lectures, ostensibly because their presence might offend the sensibilities of white members, some of whom came from the segregated south. But the policy, which lasted through the 1940s, was clearly based on his own racism.[78]

Grosvenor's preference for a hierarchical social order played itself out in the hiring practices and dining arrangements at the National Geographic Society; as the staff expanded, the Society provided separate dining rooms for the executives, the lower and middle-ranked male staff, and the female staff. The women's dining room was further segregated by occupation: businesswomen, secretaries and library and research staff, many of whom were hired as young college graduates or debutantes and not expected to stay past marriage. The separate cafeterias were supposed to diminish interactions between the unmarried women and the married men and to

---

75 "I've registered as a Democrat for the first time in my life, because I believe everyone should uphold President Wilson and Secretary Bryan. Things in a business way are picking up in the U.S.," Grosvenor wrote to his mother-in-law in 1914. (GHG to Mabel Hubbard Bell, 28 September 1914, Box 52, Bell Papers.) In a biographical sketch of Grosvenor dated 1960, Maynard Owen Williams lists Grosvenor as a Presbyterian and a Republican.

76 Bryan, *100 Years*; GHG to Jessie Holton, Principal, Holton Arms School, 11 November 1913, Box 267, Bell Papers.

77 Unidentified National Geographic employee to Maynard Owen Williams, 27 November 1944, Item 42, Box 10, Williams Collection. "Mr. H." may have been assistant editor Jesse R. Hildebrand, the editor in charge of legends (captions).

78 Poole, *Explorers House*, 63, 201–2.

protect the women from potentially rude language of the male staff.[79] The orderly halls of the National Geographic Society reflected Grosvenor's ideal political world: a humming machine of productivity where people at all levels were happy with their positions. Such a world view was precisely the one the *National Geographic Magazine* developed and promoted until well into the twentieth century.[80]

The prospect of a radical new social order dismayed Grosvenor, especially when they threatened the comfortable and powerful position of the wealthy and white. Reflections on the post-World War I negotiations in Paris moved Grosvenor to state that "if this doctrine [of self-determination … announced by President Wilson] was followed to its logical conclusion, it would mean the break-up of the British Empire. It would justify the establishment of a black republic in the United States, and it seems to me it would also justify the secession of our Southern States. The doctrine has been used to incite rebellions in Ireland, Egypt and India and I have felt [it] is one of the principal reasons of the present turmoil throughout the world."[81]

As late as the 1940s, *National Geographic* articles consistently depicted the developments of Western powers in colonial possessions as achievements to be admired.[82] In a 1930 article on Sumatra, for example, W. Robert Moore, who joined the National Geographic in 1927 and retired as chief of foreign staff in the 1960s, reported that "thanks to the efforts of the Dutch government and the missionaries, the trip to Toba Lake and its surrounding territory can be made in perfect safety," as the indigenous inhabitants were "no longer hostile to the white man."[83]

National Geographic's enthusiasm for strong governments that imposed social order in the name of national productivity extended to articles on fascist Italy and Germany. "For nearly ten years before the outbreak of World War II, it appears that Grosvenor couldn't get enough of Fascist Italy," remarks Abramson, who details the disturbing sunniness with which authors John Patric and Douglas Chandler depicted Germany, Italy, and the countries that fell under their sway.[84] Chandler, who would

---

79 Abramson, *Behind America's Lens,* 148; Carolyn Bennett Patterson, *Of Lands, Legends, and Laughter: The Search for Adventure with National Geographic* (Golden, Colorado: Fulcrum Publishing, 1998), 28. Abramson, a former *Washington Post* editor cites a 17 July 1977 *Washington Post* article that claimed hiring practices at the Geographic "were always backward toward blacks and women, some say, adding that segregated dining rooms – for men, women, blacks, blue-collar workers-abounded."

80 See Rothenberg, "Voyeurs of Imperialism"; Lutz and Collins, *Reading National Geographic.*

81 GHG to AGB, 13 May 1921, Box 100, Grosvenor Papers. The letter concerned Grosvenor's favorable opinion of Robert Lansing's book *Peace Negotiations.*

82 Rothenberg, "Voyeurs of Imperialism."

83 W. Robert Moore, "Among the Hill Tribes of Sumatra," *NGM* 57 (1930), 207. He succeeded Maynard Owen Williams in that position, and like Williams, was considered one of the Geographic's "double threats" – both writer and photographer. His Kodachromes were the first the *National Geographic Magazine* published, in 1938. Charles McCarry, "Three Men Who Made the Magazine," *NG* 174 (September 1988), 287–316.

84 Abramson, *Behind America's Lens,* 169.

later spend 15 years in a U.S. prison for being a Nazi agent, had observed in a 1937 article that, "As a substitute for Scout training, German youngsters now join an institution known as the Hitler Youth organization. Its emblem is the swastika, and its wide activities and political training are enormously popular with all classes."[85]

National Geographic was able to present a sunny picture of Nazi Germany not simply because of the magazine's proclaimed policy of refraining from negativity and politics, but because Grosvenor was anti-Semitic. He specifically wanted a story on the new fascist regime. "I hope I shall live to see Hitler unite all Germans of Germany and Austria into one powerful country," Grosvenor told La Gorce in 1933, an interesting position given that he had published anti-German articles less than 20 years earlier.[86] Douglas Chandler had four more articles published after his Berlin story, each assigned to him by the *Geographic.*

Grosvenor's tilt towards fascism stemmed from his anti-Semitism as well as his hatred of communism. In fact, each loathing fed the other. "How I hate that [Jewish] race," Grosvenor wrote to his father in 1918. "They are responsible for all the Bolshevik's propaganda everywhere and Bolshevik horrors."[87] Grosvenor had traveled through Russia in 1913 and his subsequent article, while acknowledging that the "Jews in Russia have had a very hard time of it for generations," showed equal sympathy for the Russian government. In his account of Russia's "Jewish problem," Grosvenor detailed the government's restrictions on Jews, and explained that an "alien race prospering where the native race goes hungry naturally arouses bitterness" and that the extra taxes and education and career limitations and extra taxes were measures to protect the "lethargic Slav" from "the wide-awake Jew."[88]

After the 1917 revolution, however, Grosvenor found little to sympathize with in the policies of the new Russian government, or in anything that smacked of communism. Gilbert Grosvenor "is as deep a reactionary as they come," wrote journalist George Seldes in 1943. According to Seldes, Grosvenor broke his friendship with the equally conservative DeWitt Wallace, owner and publisher of *Reader's Digest*, when *Reader's Digest* reprinted a *Geographic* article about a South Seas agricultural cooperative. The *National Geographic* article played down the part "where they grow potted palm seeds and everyone shares in the proceeds fairly and equally, according to the time worked," while *Reader's Digest* expanded on the

---

85 Douglas Chandler, "Changing Berlin," *NGM* 71 (February 1937), 170; Abramson, *Behind America's Lens*, 161.

86 Poole, *Explorers House*, 175.

87 Quoted in Poole, 118.

88 Gilbert H. Grosvenor, "Young Russia: The Land of Unlimited Possibilities," *NGM* 26 (November 1914), 486, 503. The expansively illustrated article, which takes up the entire issue, spends several pages on the corpse-laden history of the country, providing a narrative of bold Christians whose nearly 300 years under the Tatars bred in them "the hatred of Mohommedan rule" (p. 431). Grosvenor acknowledged Russian censorship, high illiteracy rate and autocratic rule, but provided no rationale for maintaining such policies, noting that "other writers have written with needless emphasis and length on these unpleasant themes, and it is not necessary to discuss them here" (p. 520).

theme. This made the *National Geographic*, which received full attribution in the digested reprint, look like a supporter of cooperative efforts. "Grosvenor is so upper-class conscious that when one of his writers sent him a piece which was mostly a travelogue on Nova Scotia, but which included a favorable mention of the great Nova Scotian cooperatives, he was fired."[89]

Seldes suggested that Grosvenor's fear of Bolsheviks stemmed from his significant stockholdings of AT&T; Grosvenor, wrote Seldes, "is scared to death about possible public ownership of public utilities."[90] Indeed, Grosvenor was an active investor, whose investments went well beyond stock in the family-derived company. His business acumen helped support his burgeoning family in a very comfortable lifestyle. He and Elsie's six (five surviving) children had nursemaids and private school educations.[91] Grosvenor kept a keen eye on the stock market, even advising his wealthy mother-in-law on stock investments.[92] In 1919, for example, during a period of U.S. military intervention in Nicaragua, Grosvenor recommended to Mabel Bell that she invest in the current strong stock of United Fruit.[93] Grosvenor took great pride in the financial achievements of the National Geographic Society: "The success of our Magazine depends on the personal attention I give it and on the business direction by me of the Society."[94] By 1920, Grosvenor could report that the Society "has today a million and a half dollars invested in real estate and bonds, another hundred thousand dollars in photographs and machinery, and the goodwill and reputation of the magazine as a going concern on a conservative valuation is worth two-and-a-half million more, making the entire property of the Society today worth approximately four million dollars."[95]

Grosvenor's hatred of communism played itself out in *National Geographic*'s coverage of Russia and the Soviet Union. As Franklin K. Lane, the former U.S. Secretary of the Interior, remarked in a 1920 article extolling the natural wonders of the United States, "The way to stand off Bolshevism is not to talk about it; it is to do

---

89 Quoted in Abramson, *Behind America's Lens*, 171–172, from George Seldes, "*Geographic* Published U.S. Traitor, Pro-Fascist Dope of *Reader's Digest* Man," *In Fact* (4 October, 1943), 3.

90 Ibid.

91 "Our babies have come fast and I am determined that Elsie shall not lose her freshness and beauty; I want her to keep her youth and loveliness for her children to admire and appreciate when they grow up, as my father kept my mother young. He spent his money on nurses instead of saving it and we boys are proud that he did it… . We have a lot of nurses, but we don't leave the children entirely to them as you think. Elsie is relieved of the drudgery of bathing them and keeping them clean, but she gives them an immense amount of intelligent care and teaching and I also try to do my part." GHG to Mabel Hubbard Bell, 13 February 1909, Box 52, Bell Papers.

92 Abramson, *Behind America's Lens*, 172; GHG to Mabel Hubbard Bell, 6 May 1919, Box 52, Bell Papers.

93 GHG to Mabel Hubbard Bell, 6 May 1919, Box 52, Bell Papers.

94 GHG to Mabel Hubbard Bell, 13 February 1909, Box 52, Bell Papers.

95 GHG to AGB, 23 April 1920, Box 100, Grosvenor Papers.

things which show … that this land is the best of all lands."[96] *National Geographic* proceeded to stand off Bolshevism in just such a manner, publishing very few articles about the Soviet Union on the one hand and promoting the United States as the best of all possible worlds on the other.

Staff writer-photographer Maynard Owen Williams knew that the Soviet Union was a touchy matter for the *Geographic*. "I do know and appreciate the burden that weighs on the responsible editor when he deals with a subject about which feeling runs high," he wrote to Grosvenor in October 1926. "I think the Geographic is doing a brave thing in publishing an article on Soviet Russia now, and there may be a flare-back, even from an unobjectionable article." Williams had furnished photographs for the piece by Junius B. Wood titled "Russia of the Hour: Giant Battle Ground for Theories of Economy, Society, and Politics, as Observed by an Unbiased Correspondent." He also urged Grosvenor to publish an article that would temper a highly negative piece by Thomas Whittemore published in 1918, a "bitter attack touching on Bolshevik spoilation of the churches." Williams had talked to a Soviet Press Bureau official "and he characterized that article as one which went outside the realm of science and geography to attack deliberately the Bolsheviks."[97] Williams wanted to get back into the Soviet Union to "make a thorough survey of some of the Soviet's little-known races." He wrote to the Press Bureau official "asking him what encouragement he can give me to re-enter Russia, and assuring him that under no circumstances can we bind ourselves to print anything favorable but that after weeks of study I am personally very friendly toward the Soviet experiment within the U.S.S.R. boundaries." Reassuring Grosvenor, Williams added that "Naturally I have not in any way committed The Society to any such plan. I have not said that even I would return to the Soviet Union."[98]

It probably did not help Williams' request for another Soviet Union assignment that he told Grosvenor that "Crusading for the Soviet is not my job, even if it were my

---

96 Franklin K. Lane, "A Mind's Eye Map of America," *NGM* 36 (June 1920), 510. While the article's many illustrations make it seem simply an opportunity to print photographs of picturesque scenery, the text itself is a call for the Americanization of immigrants through such means as improving teachers' salaries and getting the American film industry to depict appropriate characters and themes. "You must interpret to them in terms of American life – the beauty of American life, its dignity, the generosity of our natures, our willingness to be fair, our desire to help, our knight-like qualities."

97 MOW to GHG, 12 October 1926, National Geographic Society Records Library (hereafter referred to as NGS Records Library). The articles Williams referred to are Junius B. Wood, "Russia of the Hour: Giant Battle Ground for Theories of Economy, Society, and Politics, as Observed by an Unbiased Correspondent," *NGM* 50 (November 1926): 519-598 and Thomas Whittemore, "The Rebirth of Religion in Russia: The Church Reorganized while Bolshevik Cannon Spread Destruction in the Nation's Holy of Holies," *NGM* 34 (November 1918): 379-401.

98 MOW to GHG, 12 October 1926; MOW to GHG, 26 October 1926, NGS Records Library.

inclination."[99] Despite his suggestion that "One-sixth the land surface of the globe – unhackneyed material – live popular interest after a long dearth of articles – Russia looks like a big bet to me,"[100] Williams did not get to do the article he requested. As he reminisced to Grosvenor after they had both retired, Grosvenor had told Williams "something like 'As long as I am editor' (and that made it eternal to me) 'you will not do an article on Russia for the N.G.M.'"[101] Still, Williams' 1942 "Mother Volga Defends Her Own," was among seven articles and a map that appeared during the brief period in World War II when the Soviet Union, the U.S. and Great Britain were allies. Indeed, Wood's 1926 article was the last on the Soviet Union until a 1932 article that described a scientific journey into Asian Russia.[102] Lest any article on the land of the Bolsheviks appear while Roosevelt was leading the U.S. through the New Deal, Grosvenor published only a single story, in 1937, about an eclipse in the far eastern part of the Soviet Union.[103]

The rationale given for *National Geographic*'s sunny coverage of places in which terrible things were happening – or the avoidance of such places – was embedded in its "guiding principles," published in the magazine in 1915:

> 1. The first principle is accurate accuracy. Nothing must be printed which is not strictly according to fact. …
> 2. Abundance of beautiful, instructive, and artistic illustrations.
> 3. Everything printed in the Magazine must have permanent value, and be so planned that each magazine will be as valuable and pertinent one year or five years after publication as it is on the day of publication. The result of this principle is that tens of thousands of back numbers of the magazine are continually used in school-rooms.
> 4. All personalities and notes of a trivial character are avoided.
> 5. Nothing of a partisan or controversial character is printed.

---

99 MOW to GHG, 26 October 1926, NGS Records Library. Williams' liberalism, as well as his earnest internationalism (see Chapter 4), gave him a perspective on the Soviet Union that was anathema to Grosvenor and many others. Referring to Harrison Salisbury's *New York Times* articles on the post-Stalin U.S.S.R., Williams argued that they "make far more sense than that American self-hypnotism entitled 'We refuse to live on the same earth as Communism.'" He later commented that "The brass-tacks folks have us shivering in our sleep, which is bad, or ready to wipe up Russian kids from so far off that no scream, danger or memory need result, which is worse." (7 October 1954 and 24 October 1954, MOW to Weimer K. Hicks, Trustees Folder "MOW 1953–55," Weimer K. Hicks Papers, Kalamazoo College, Kalamazoo, Michigan).

100 MOW to GHG, 12 October 1926, NGS Records Library.

101 MOW to GHG, 12 March 1961, Folder "Grosvenor bio notes," Item 29, Box 7, Williams Collection.

102 Lyman D. Wilbur, "Surveying Through Khoresm: A Journey into Parts of Asiatic Russia Which Have Been Closed to Western Travelers Since the World War," *NGM* 61 (June 1932): 753–780.

103 No articles were published on the Soviet Union from 1946 until 1959 – and then only with a major preface to offset any possible misconception that the Society did not consider the Soviet Union to be a political enemy of the U.S. See Abramson, *Behind America's Lens*, 194–5.

6. Only what is of a kindly nature is printed about any country or people, everything unpleasant or unduly critical being avoided.
7. The contents of each number is planned with a view of being timely.[104]

The principles of accuracy, timeliness and abundance of illustration had been developed in the earliest days of Grosvenor's employment, under the tutelage of Alexander Graham Bell. Given Bell's frequent advice to Grosvenor to cover the latest global happenings – he even suggested a department of the magazine "devoted to the geography of current events"[105] – it is most likely Grosvenor who pressed time*less*ness along with time*li*ness. The axiom of timelessness, set down in principles three, four and five, expresses the magazine's commitment to portray the "essential nature" of places and peoples. How *National Geographic* set about, consciously and unconsciously, to perpetuate the concept of national, ethnic, and racial "types," will be discussed in the following chapter.

But the societies and places covered by the *Geographic* were neither timeless nor free of political tensions. It is not that *National Geographic* ignored political structures or personalities. There were many articles that featured kings, princes, and chiefs, as well as colonial administrators and governors. What the magazine pointedly omitted were depictions of political struggle or analyses of political or political-economic circumstances. The effect of such omission was to lend legitimacy to the reported status quo.

When *National Geographic* did venture into political territory, it was firmly from the perspective of the imperial powers, particularly that of the U.S. In 1920, *The Nation* magazine took *National Geographic* to task for portraying the American military takeover of Haiti, initiated in 1915 under President Woodrow Wilson, as a positive development.[106] An unbylined article called "Haiti and Its Regeneration by the United States" presented a distinctly political historical overview of the country, from French colonial days to the present. The article made several parallels between the U.S. involvement in Haiti and in the Philippines, with the understanding that the U.S. had improved the Philippines' health, education, infrastructure, morality and political development. "As the campaign in the Philippines proved a success, there is every reason to believe that within a similar period equal progress could be made in the Haitian Republic and within twenty years we should have a people speaking English almost as universally as Haitian."[107]

The Nation reprinted an exchange between a reader and La Gorce regarding the *Geographic* article, with additional commentary. La Gorce defended the accuracy

---

104 Grosvenor, "Report of the Director and Editor of the National Geographic Society for the Year 1914," *NGM* 27 (March 1915), 319.

105 AGB to GHG, 7 December 1905, Box 267, Bell Papers.

106 "Our Imperialist Propaganda: The *National Geographic*'s Anti-Haitian Campaign," *The Nation* 112 (6 April 1921). http://www.boondocksnet.com/ai/ailtexts/nation210406.html, in Jim Zwick, ed., *Anti-Imperialism in the United States, 1898-1935*, http://www.boondocksnet.com/ai (accessed 26 February 2004).

107 "Haiti and Its Regeneration by the United States," *NGM* 38 (December 1920), 511.

of the article, noting that an author of a recent book on the West Indies as well as several members of the National Geographic Society "who have had an opportunity to study conditions in that country" had examined the piece. Despite the fact that *The Nation* and *The New Republic*, among other publications, had reported that U.S. Marines had pressed Haitians into forced labor and killed thousands, La Gorce reiterated that the *Geographic*'s portrayal valorizing the Marines was appropriate and accurate, and that the Marines had faced not just chaos, but guerilla warfare. "The same condition that occurred in the Philippines when we were forced to take them over … occurs in a smaller way in Haiti, and requires drastic action, and this only as brought about a house-cleaning."[108] La Gorce refuted the idea "that the *National Geographic Magazine* could be subsidized for propaganda of any kind or description, for its honorable record of thirty-two years certainly precludes any such possibility."[109] This public dispute, however, suggests that despite the *Geographic*'s success in maintaining a reputation for accuracy, if not objectivity, not everyone accepted that estimation.

The two world wars tested the principles of non-partisanship and avoidance of controversy, but there was never any question that the National Geographic Society would do whatever it could to support *its* national government. In fact, the Society provided valuable cartographic assistance in both wars.[110] Grosvenor, who had been in Europe in 1913, heard rumblings of unease and decided upon his return to have an updated map of central Europe prepared; as soon as war was declared in 1914, copies were published as a supplement to National Geographic Society member-subscribers.[111] The Geographic also initiated daily newspaper bulletins during the First World War, providing background about overseas battle zones and, in the case of bulletin writer Harriet Chalmers Adams, news from the front as well. The Geographic offered the bulletins free of charge to newspapers and in return, the Geographic got free publicity and an enhanced reputation as a source of accurate information.[112]

As soon as the war began, the *Geographic* started a series of encyclopedic articles on the countries involved in both sides of the war. Then the magazine devoted its attention to other parts of the world, particularly the Americas. But *National Geographic* jumped in on the Allied side even before the U.S. declared war on Germany. Starting with the March 1917 issue, the magazine went full-throttle with patriotic pieces on the heroism of the military of the Allied nations and the sacrifices of their citizens. *National Geographic*'s coverage of the war, including its maps and its newspaper bulletins, contributed to the doubling of the magazine's circulation,

---

108 "Our Imperialist Propaganda," 2. Regarding the U.S. being "forced" to take over the Philippines, *The Nation* commented that La Gorce seemed "to be unaware of the war with Spain" (p. 4).

109 Ibid.

110 Grosvenor, *NGS and Its Magazine*; Schulten, *Geographical Imagination in America*.

111 Bryan, 131.

112 Abramson, *Behind America's Lens*, 182.

from 337,000 in 1914 to over 625,000 in 1917. Susan Schulten notes, however, that the Society's growth actually stagnated once the U.S. joined the war, and suggests that it was the explicit political bias that turned members, an increasingly varied group, away.[113] She cites angry letters from readers criticizing the *Geographic* for sinking "to the low place of a common political magazine," printing "bombastic slush" and losing its "scientific impartiality."[114] Still, with thousands of men headed off to war, it would have been the first time that so many members might not have been around to renew their memberships.[115]

"The *Geographic* always dealt in facts, not bias, rumor, or prejudice,"[116] Grosvenor reiterated toward the end of his life. And perhaps he did consider it as a fact, not a political perspective, that as of 1927 the "United States has repeatedly endeavored to bring peace out of the Nicaraguan chaos."[117] Perhaps he did not consider it prejudice or rumor to say that the low-caste "boy" a traveler might hire in Rangoon as a servant "is usually a worthless, no-account fellow, whom no resident, white or native, will employ."[118] Or that at a train station in Egypt, "Voluble Arabs made the usual din, apparently about nothing."[119] The point is not that every *National Geographic* article smoothed over or endorsed imperialist activities or was laced with condescending attitudes, but that these forms of narrative regularly appeared at the same time that the magazine was representing itself as an objective, educational source of pure geographical information. And these were the "facts" that Americans were learning as they absorbed the popular geography of the *National Geographic Magazine*.

Gilbert Grosvenor retired as president of the National Geographic Society and editor-in-chief of the magazine in 1954, at the age of 78. He was succeeded briefly by La Gorce, who then also retired. The reins were handed over to Melville Bell Grosvenor, the Chief's oldest child, who was by then 55. The extent to which Gilbert Grosvenor's reactionary politics colored his definitions of "political" and "apolitical" and framed his dedication to printing "only kindly" things about a place can be seen in a letter he sent his son Melville in 1960 regarding a *National Geographic* article on South Africa. "There are several unpleasant references to 'apartheid' on pages 321, 325, 347," he complained to Mel. "Of course none of us approve of 'apartheid' but every member of the NGS reads about that terrible situation *daily* in their newspaper and every week in 'Time', 'Newsweek' etc and every month in 'World Report', 'Atlantic Monthly'. Why therefore give space to describe 'apartheid' in the NGM." The elder Grosvenor resented what he saw as a revisionist take on African colonialism: "Today it is the fashion for writers to represent the white man as the enslaver of Africa and I am afraid Mr Kenney has followed that pattern," he wrote,

113 Schulten, *Geographical Imagination in America*, 151.

114 Ibid., 168.

115 Bryan, *100 Years*, 135.

116 Grosvenor, "Story of the Geographic," 44.

117 "Nicaragua, Largest of Central American Republics," *NGM* 51 (March 1927), 378.

118 Charles H. Bartlett, "Untoured Burma," *NGM* 24 (July 1913), 835.

119 Felix Shay, "Cairo to Cape Town, Overland," *NGM* 47 (February 1925), 131.

calling the author's presentation "a distortion of the white man's work in Africa. *The white man, with missionaries, engineers, schools, awakened Africa* and now the white man is being driven out by the men he educated and Africa will soon become again the Dark Continent," Grosvenor continued. "I am proud of being a white man. I am sorry to see this glorious work in Africa misrepresented."[120]

Despite the fact that Gilbert Grosvenor had retired six years earlier, he still held enormous sway. The original language of the article read: "'Men with pale skins came from over the seas. Finding Mother Africa drugged with sleep, they made her their servant. But bonds rattle – The clangor of her fetters, awakened Mother Africa.'" The actual published article metamorphosed "men with pale skins" into specialized "slavers" who "seized her children and hurried them away." Mother Africa – the singular feminized entity – then encounters European explorers followed by "Hosts of missionaries, traders, engineers, doctors and teachers… . Devoted scientists came to fight the tsetse fly that devastated half a continent."[121] Not a single colonist per se, or European soldier, or administrator, or resource-extracting company is mentioned. The word "apartheid" is also absent from the published article. There are only allusions to apartheid but no actual discussion of apartheid policies or apartheid's geography.[122]

In Gilbert Grosvenor's world – and thus the world of *National Geographic* – it would be a "misrepresentation" to suggest that white Europeans took the African continent captive. However, it apparently would not be a "misrepresentation" to say that "African natives are simple, child-like creatures, whose symbolism is as primitive as their other instincts,"[123] a statement which appeared in the magazine as late as 1938. Grosvenor understood the power of representation. He intended the *National Geographic Magazine* to be a magazine that would show Americans the world that is "ours," *ours* as earthlings all together, but at the same time, *ours* as Americans, his *us* specifically white middle- and upper-class Americans. As Grosvenor relayed to the Association of American Geographers in the mid-1920s, more than a million copies of the magazine "go to every community in the United States of 50 or more white persons, and to 152 nations, colonies and mandatories;

---

120 GHG to Melville Bell Grosvenor, 23 June 1960, Box 140, Grosvenor Papers. The article is Nathaniel T. Kenney, "The Winds of Freedom Stir a Continent," *NGM* 118 (September 1960): 303–359; the italics (underlining in the original) is Grosvenor's.

121 Kenney, "Winds of Freedom," 303.

122 The author juxtaposes black labor with white wealth, and sympathetically notes a Zulu porter's comment that he cannot enter a park in his own country even though the foreign white writer can. The comment is immediately followed by a Boer farmer stating he'd rather burn his home than hand it over to blacks, with the two "sides" mediated by a third comment from "a cultured man" of Cape Town, who states that the Zulu and the Boer would like each other if they could sit in the park and talk to each other, but that "There is no solution to our problem." Kenney, "Winds of Freedom," 325–326.

123 Lawrence Thaw and Margaret Thaw, "Trans-Africa Safari," *NGM* 74 (September 1938), 342.

to every country, in fact, which has a postal system."[124] Although proud that it was delivered to corners all over the world, *National Geographic* chose to ignore what it did not like. It imagined for itself and its readers a nation of "white persons." The denial of "Americanness" for African-Americans, and their exclusion from *National Geographic*'s imagined nation of readers can be seen in that same 1938 article referred to above. In it, the authors compare West African habits to those of "our American negro."[125] Unaddressed as part of the *National Geographic* reader-citizenship, African-Americans are referred to only in possessive relation to the white American reader, the assumed "us."

It was to this imagined community of a white, propertied, and Anglo-Christian-assimilated nation that Grosvenor styled the Geographic; not only was this the nation he wanted to serve, this was the nation he wanted. Anything deemed threatening to this nation – communism, anti-colonization struggles, American poverty, African-Americans, unassimilating immigrants – would not and did not exist in *National Geographic*'s world. And yet, at the same time, the magazine trumpeted its purity of intent and its objectivity. "The success of the magazine must always rest primarily on two factors," the first being "retaining the confidence of the public in the absolute accuracy and impartiality of its contents."[126] Whatever the stated intentions of objectivity and truthfulness, the *National Geographic Magazine* depicted, to increasing numbers of readers, a world molded in its own imagining. In this world, the U.S. was modern and paternalistic, and Europe was cultured but quaint. As for the rest of humanity, it resided in various stages of faded glory, historical marginality, apprenticeship to civilization, or childish revelry.

Ever since Henry Gannett simultaneously headed both the U.S. Board of Geographic Names *and* the National Geographic Society,[127] rendering the *National Geographic Magazine* the authoritative source for geographical place names, the Society benefited from the public's confusion regarding its status in relation to the United States government. National Geographic identified with a particular vision of the United States and it established itself as a "circulation leader" by selling that vision to the American public. The representational forms of that vision – in particular, photography – are the subject of the next chapter.

---

124 From notes for Gilbert H. Grosvenor, "Some Recent Work of the National Geographic Society" 21st annual meeting of the Association of American Geographers, Washington, D.C., 30 December 1924 – 1 January 1925, Folder NGS early materials, Box 159, Grosvenor Papers.

125 Thaw and Thaw, "Trans-Africa Safari," 342.

126 "Report of the Director, 1919." The second factor was "giving the public interesting geographic information on those subjects to which their attention is directed by international events."

127 Gannett helped organize the Board of Geographic Names, and served as chairman from 1894 until his death in 1914. An original member of the NGS Board of Managers, Gannett served briefly as editor of the magazine (between Hyde and Grosvenor), and was president of the Society from 1910 until his death. See Lacey,"Earth-Making Dissolve"; Grosvenor, *NGS and Its Magazine*.

Chapter 3

Picturing Human Geography: Orders of Science and Art

In August 1995, *National Geographic* inaugurated its "Flashback" feature, a back-of-the-book item singling out a notable photograph from the National Geographic Society collection. Ninety years earlier, Gilbert H. Grosvenor had gambled on publishing copious photographs of Lhasa and the Philippines, and won, netting positive attention for the magazine and a bigger membership for the Society, and steering the magazine toward more photography. By 1995, the Geographic's archives were filled with hundreds of thousands of photographs, a rich resource ready to be mined. The Flashback feature enabled the *Geographic* to showcase some of these photographs, some of which had never been published, and made all the more interesting because they presented not just a window on the world, but a window onto the past.

National Geographic's November 1996 Flashback featured a previously unpublished photograph from the early twentieth century of "a Marquesan chief, in a cape of human hair taken from the enemy dead, and his companions." In keeping with the *National Geographic* tradition of deflection, the caption did not directly mention the fact that the man's young female companions were naked. Rather, it repeated the story regarding the origins of the magazine's policy on "native" nudity: "When Editor Gilbert H. Grosvenor published photographs of unclad Philippine women in 1903, he said the pictures were 'a true reflection of the customs of the times in those islands.'" Grosvenor's statement functioned to demonstrate the supposedly "objective" rationale for publishing photos of "unclad women," exculpating the present editors from making that claim themselves. The fact that the photograph was carefully posed, or that the image of an older man surrounded by naked nubiles conformed to Western male sexual fantasies of life in the Pacific islands, went unmentioned.[1]

Instead of commenting on or deconstructing this durable Western fantasy in its November 1996 Flashback, *National Geographic* resorted to backhanded reasons for the photograph's inclusion. The photograph's caption referred readers to that issue's feature story on Joseph Banks, the British botanist who accompanied Captain Cook in his first South Pacific voyage in 1768, and then denied any real connection of the photograph to the story. As the caption, headed "What Joseph Banks Missed,"

1 *NGM* 190 (November 1996), 130. The photograph is credited to L. Gauthier, photographer of the young Marquesan woman on p. 91.

explained, not only had Banks never visited the Marquesa Islands, but the photograph was not selected for the 1919 article "A Vanishing People of the South Seas."

This long-term consistency in *National Geographic*'s presentation of erotic exotica was brought into high relief by a nearly concurrent display of early South Pacific photographs that took a far more critical approach. During the summer of 1996, the Metropolitan Museum of Art in New York City mounted a small exhibition of colonial photography of Samoa. Alongside the images of palm-framed vistas and long-haired women, the accompanying text made its revisionist intent clear. In keeping with late twentieth-century cultural and post-colonial studies sensibilities, "Picturing Paradise: Colonial Photography of Samoa, 1875–1925" took note of the political and economic relationships between the islands and the various Western countries that exerted dominance over Samoa, guiding the museum-goer into seeing how these photographs built on, added to, and reformulated the Western image of Samoa as Pacific Island Paradise. The photographs themselves ranged from carefully crafted studio portraits of Samoans and lush landscape photographs largely for commercial use (as book illustrations or as tourist postcards), to stereoscopic views for armchair travelers, to personal photographs of Western visitors and anthropologists, to movie production stills. The accompanying text noted that these photographs offered information and insight into a world now decades away, and might be used, with judicious caution, as historical documents of Samoan life. But, the commentary argued, perhaps the greater significance of these photographs, and a more legitimate way of understanding them, was as documents of Western ideas of Pacific paradise.[2]

The fact that in 1996 *National Geographic* could still uncritically present a photograph in a manner consistent with its early twentieth-century approach suggests that a historical examination of the magazine's photographs still has contemporary relevance. It is clear that *National Geographic* chose not to interrogate its past representations or raise the issue of the conflicting knowledges such photographs present. In contrast, the "Picturing Paradise" exhibit contextualized the photographs, thereby widening the possible answers to the question, "What do the photographs tell us about Samoa in the late nineteenth and early twentieth century?" How is that potential knowledge of "fact" modified, manipulated, constructed or obliterated by the context – and by our recognition of the context – in which these photographs were taken? How does one read such an entanglement of fact and fantasy, of imperialism, curiosity, and technology?

This chapter takes on a similar task of reevaluating early twentieth-century Western representations of non-Western peoples and places. As its own histories

---

2 1 March–4 August 1996. Co-produced by the Southeast Museum of Photography, Daytona Beach and the Rautenstrauch-Joest Museum of Ethnography, Cologne, with support of the office of the President, Daytona Beach Community College, and the City of Cologne. Project designated with UNESCO label for cultural significance. In New York made possible by the Friends of the Arts of Africa, Oceania, and the Americas.

proudly note, *National Geographic* built its success on photography.[3] "The National Geographic Magazine has found a new universal language which requires no deep study – … the Language of the Photograph!" declared John Oliver La Gorce in his 1915 *The Story of the Geographic*.[4] Unlike the illustrations used by other magazines, La Gorce continued, the *Geographic* offered "Talking Pictures" that told their own story, so that "knowledge is planted without the reasoning process being unduly taxed or by subsequent disturbances of the mental digestive track in the assimilation of new facts."[5]

Like most new languages, however, the "language of the photograph" was built from an amalgam of older languages; its grammar and meanings were familiar to those already fluent in the earlier languages. While La Gorce, writing in the early twentieth century, could extol the obvious simplicity and transparency involved in "reading" photographs, a more contemporary understanding of the link between photography and imperialism complicates the undertaking. As George Stocking notes, the "colonial situations" in which the photographs were made are "an essential context and a legitimate goal of interpretation."[6] Thus photographs "do not eye-witness an unmediated reality," nor do they "reveal their meaning directly to the naked eye of the reader."[7] Taking as its premise that *National Geographic* produced an American serial version of what Mary Louise Pratt has termed "anti-conquest" narrative, this chapter focuses on the narrative device most important to the success of the magazine: photography. Photography was uniquely capable of combining two important strains of the anti-conquest narrative, "science" and "sentiment,"[8] and *National Geographic* provided a highly effective vehicle for this "modern" contradictory travel narrative.

Science and art in the exploration and visualization of the world

Geography's emergence as a modern science can be traced to Cook's first voyage in 1768. It was then that, as Derek Gregory notes, Cook's "scientists, collectors and illustrators displayed three features of decisive significance for the formation of geography as a distinctively modern, avowedly 'objective' science: a concern

3 La Gorce, *Story of the Geographic*; Grosvenor, *The NGS and Its Magazine*; Bryan, *100 Years*.

4 La Gorce called the language of the photograph one which "takes precedence over Esperanto and one that is understood as well by the jungaleer as by the courtier; by the Eskimo as by the wild man from Borneo; by the child in the playroom as by the professor in the college; and by the woman of the household as well as by the hurried businessman."

5 La Gorce, *The Story of the Geographic*, 3.

6 George W. Stocking, Jr., "The Camera Eye as I Witness: Skeptical Reflections on the 'Hidden Messages' of *Anthropology and Photography, 1860–1920*," *Visual Anthropology* 6 (1993), 218.

7 Stocking, "Camera Eye," 215.

8 Pratt, *Imperial Eyes*.

for realism in description, for systematic classification in collection, and for the comparative method in explanation."[9] David Stoddart has argued that it is "the extension of scientific methods of observation, classification and comparison to peoples and societies" that made the subject of geography possible.[10] Geography's position as a modern science, then, is inherently bound with its development as a social science.[11]

In launching his history of modern geography with Cook, Stoddart unabashedly reinforces the association of the discipline with exploratory travel. This linkage, as Gregory notes, conflates exploration with adventure – adventure as a romantic notion and activity.[12] Gregory uses Stoddart's romantic imagery to support his own

---

9 Derek Gregory, *Geographical Imaginations* (Cambridge, MA: Blackwell, 1994), 19.

10 D.R. Stoddart, *On Geography and Its History* (New York: Basil Blackwell, 1986), 35. Oddly, in the same paragraph Stoddart relays an anecdote from Bougainville's 1772 *A Voyage Round the World* in which Bougainville's ship, upon arrival in Tahiti in 1768, "was surrounded by canoes of naked women as it edged through the reef, and the marines had to be called out to keep order among the sailors." A cook who went ashore was immediately besieged by the women, who had their way with him. Although Stoddart sets up a sympathetic explanation for the picturing of Pacific Islands as fantastic idylls – the voyages from Europe were long and nasty – his use of Bougainville's Tahitian story, in which naked women inhabitants are both enticing and terrifyingly aggressive, serves as well (regardless of Stoddart's intentions) to suggest how intrinsically racialized and gendered these "scientific methods of observation, classification and comparison" were.

11 While Stoddart touts Cook's voyage for its empirical contributions, Bernard Smith, *European Vision,* finds the same expedition significant for its artwork, drawings of Pacific landscapes and peoples for European audiences, produced by Sydney Parkinson and Alexander Buchan under the direction of botanist Joseph Banks.

12 The *National Geographic* November 1996 article on Joseph Banks heralds him as one of history's greatest botanists and in the process links exploration-as-romance, the imperialist intentions of European naturalist science, and the magazine's own honored tradition of using claims of "truth" to publish photographs of bare-breasted brown women. (T. H. Watkins, "The Greening of the Empire: Sir Joseph Banks, *NG* 190 (November 1996), 28–52). The article sets the stage by introducing Banks as a dashing young wealthy aristocrat, "full of the juices of life, … with dark liquid eyes and a mouth that a romance novelist of today would describe as sensuous" (p. 36). With comments such as Banks's voice "boom[ing] with good fellowship" (p. 36), his "talent for the social arts" (p. 36), and the suggestion that Banks "had to have been positively drunk on exploration" (p. 42), author T. H. Watkins successfully intertwines Banks-the-scientist with Banks-the-romantic-hero, playing out the trope of exploration as romance. The portrait of a vibrant young Banks contrasts with the "overbearing and imperious man" described by Stoddart, who "personified for four decades hegemony in practice" (Stoddart, *On Geography*, 18–19).

The *National Geographic* story also explicitly spells out Banks's role as an imperialist. Promoting the settlement of British convicts in Australia's Botany Bay, recommending the transplanting of Tahitian breadfruit to the West Indies for the purposes of feeding Britain's African slaves, and devoting himself to obtaining the most plant species in the world for King George III's Royal Botanical Gardens, Banks "practiced a brand of biological imperialism, and his collectors were ordered to be alert to how the medical or economic potential of what

argument that geography's "world-as-picture" framework – recognizable in the early twentieth-century geographers Vidal de la Blache and Carl Sauer as much as in the early nineteenth-century geographers Humboldt and Ritter – has within it a sensibility "at once aesthetic and scientific."[13] Humboldt, for example, whom Pratt establishes as a classic example of an anti-conquest writer,[14] consciously aimed to unite science and art in his work, "where the sensuous and the objective existed side by side, sometimes in a single sentence."[15] Barbara Stafford, in her study of eighteenth-century "scientific explorer-artist-writers" argues as well that "the study of natural history," which she shows to be intimately bound with scientific travel, "helped to undermine the strict Baconian tradition that established an antithesis between science and poetry, thinking and feeling."[16]

This balance of science and aesthetics – sometimes configured as science and art, sometimes as science and romance, sometimes as science and sentiment – surfaces repeatedly in analyses of exploration and of photography.[17] While the terms and concepts opposed to and in concert with science – art, aesthetics, romance, sentiment – have different inflections and intellectual histories, in each pairing, science stands for methodical, rational, objective knowledge. Science's contradictory partner, in each case, is primarily concerned with the subjective, and with feeling.[18] Mary Louise Pratt, for example, sets out the two most important anti-conquest narrative strategies as those of "science and sentiment."[19] While the language of science is that of objectivity, unambiguousness and pure information, the complementary language of sentiment is that of "desire, sex, spirituality, and the Individual."[20] Each works

---

they found might further British interests" (p. 51). Knowledge and conquest are explicitly linked in Banks's life (as well as in the *National Geographic* story of Banks's life), feeding nicely Pratt's argument for the critical role of naturalists in the European planetary project.

13 Gregory, *Geographical Imaginations*, 40.

14 Pratt, *Imperial Eyes*.

15 Edmunds Bunkse, "Humboldt and an Aesthetic Tradition in Geography," *Geographical Review* 71 (1981), 137. Bunkse discusses Humboldt's theory of the symbiotic relationship between objective science and subjective "awe" of nature. Humans' "'mysterious' apprehension of a universal order in nature" leads them to ask questions regarding nature's formations, use reason to seek answers to these questions, and design instruments to aid their inquiry. Science, in turn, "furthers the intuitive, subjective enjoyment of nature," the source of continual inspiration (p. 138). Note as well how "Nature" in this schema is profoundly feminized, acting as both muse and object of study.

16 Barbara Maria Stafford, *Voyage into Substance: Art, Science, Nature, and the Illustrated Travel Account, 1760–1840* (Cambridge, MA: The MIT Press, 1984), 2, 55.

17 See for example, Ryan, *Picturing Empire*.

18 The early twentieth-century Italian philosopher Benedetto Croce, for example, held that "Art is a true *aesthetic a priori synthesis* of feeling and image within intuition," and that feeling and image require "the aesthetic synthesis" to render them as art. Benedetto Croce, *Guide to Aesthetics*, Patrick Romanell, trans. (New York: Bobbs-Merrill, 1965). *Brevario di estetica* was originally published in 1913.

19 Pratt, *Imperial Eyes*, 39.

20 Ibid., 78.

the narrative's contradictions in different ways, the former positioning itself so as to transcend real politics in the quest for objective knowledge, the latter poised to transcend the ordinary through heightened subjective feeling.

In Donna Haraway's discussion of the American Museum of Natural History's display of formerly animated objects collected on such scientific expeditions, "artistic realism and biological science were twin brothers in the founding of the civic order of nature."[21] Haraway's "twin brothers" metaphor skillfully elicits the close relationship between art and science, as well as the masculine gendered connotations of both. One shared component of both science and art, or alternately, science and romance, is that of distance: specifically, a distance from nature. Science is based on a principle of objective detachment of the observer from the observed; nature is science's object of study. Zuleyma Tang Halpin describes objectivity as the ability "to see one's self as totally different and apart from the object of the study, the 'other.'"[22] Traditional Western representational art requests a similar distancing of the artist from the subject, a concept elucidated in discussions differentiating the naked (a "cultural" condition of shame and sexuality), which has no place in "high" art, and the nude (a "natural" condition of decontextualized form); nudes are acceptable *because* they are expected to be viewed in an unemotional manner.[23] The artist is expected to maintain some degree of detachment from the source of study, no matter how familiar, or how naked, in part because the subject becomes objectified as a purely natural form. The chemist and art collector Albert C. Barnes saw another similarity between art and science: "The artist is primarily the discoverer, just as the scientist is."[24]

Both scientific and aesthetic modes were deeply embedded in narratives of exploration. "It was the voyager … who paved the way for works of nature to be validated aesthetically,"[25] Stafford argues. "It was in the explorers' implementation of a method of discovery based on a willed nonmetaphoric scrutiny of the particulars

---

21 Haraway, *Primate Visions*, 38–39. The Museum shared its ethic of nature, masculinity and republican imperialism, and many of its endeavors – including emphasis on visual representation, dedication to the popularization of science, and sponsorship of expeditions – with *National Geographic* (p. 56).

22 Zuleyma Tang Halpin, "Scientific Objectivity and the Concept of the 'Other,'" *Women's Studies International Forum* 12 (1989), 285–94. Halpin, among others, argues that the scientific idea of objectivity only exacerbates the "othering" of non-white, non-male people.

23 For example, "a truly beautiful artistic nude should have no pornographic suggestions or connotations; if it has any, the creative artist has willed it so, as self-indulgence, or as part of preoccupations – commercial or other – that are alien to art." Joseph Chiari, *Art and Knowledge*, (New York: Gordian Press, 1977), 27. See also Kenneth Clark, *The Nude: A Study in Ideal Form* (Garden City, New York: Doubleday, 1956).

24 Albert C. Barnes, *The Art in Painting*, revised edition (New York: Harcourt, Brace and Company, 1937), 13.

25 Stafford, *Voyage into Substance*, 486.

of this world that truth telling was elevated to aesthetic status."[26] Since photography had already been the arena for debates as to its status as artistic or scientific, it should not be surprising that when brought into a tradition of exploratory narrative, photography would thus play a dual role. Its record of "exotic" peoples and places could be read as at once factual and imbued with sensual, if only visual, pleasure.

Natural history, making "the order of things visible"

That historian Alan Trachtenberg chose to open his study of American photography with a painting, American artist Charles Willson Peale's 1822 self-portrait "The Artist in His Museum," is striking. The elderly Peale stands behind a dead turkey, taxidermist's tools, a large animal's leg and jaw bones, and a painter's palette. He lifts a curtain to reveal a long, sunlit room, in which we see floor-to-ceiling display cases, each filled with taxidermy specimens; the top row belongs to painted portraits.

> Following Linnaeus's designations of class and genus as the principle governing universal order, the painting illustrates the artist in his task of illuminating the unity of nature, a unity based on a coherent structure of forms rising from the simplest to the most complex – the "great chain of being" of Enlightenment thought. Thus the serial ranking of specimens in the Long Room of the museum: lower order of creatures –ducks and penguins – on the bottom, rising through songbirds and birds of prey to the row of portraits of distinguished public figures at the top. Art clarifies the system, brings it to view. To make the order of things visible requires two commensurate acts: clarity of representation, and formal placement of the represented thing within a strictly defined spatial form.[27]

As suggested by the painting, taxonomy – Linnean taxonomy in particular – is the structure around which this knowledge is to be built. Contemporaneous with Cook and Banks, Linnaeus figures equally prominently in discussions of the origins of European visions of non-Europeans. Linnaeus, an eighteenth-century Swedish naturalist, created the classification system of Latin names indicating increments

---

26 Ibid., 1.

27 Alan Trachtenberg, *Reading American Photographs: Images as History, Mathew Brady to Walker Evans* (New York: Hill and Wang, 1989), 8–9. That Peale includes himself as well as "his" workshop in the painting further points to the role of the artist as "mediator of the world's truth" (p. 9). But Peale, who had founded the museum in the painting, was considered a scientist as well, and one could well interpret his figure as standing on the soft, pliable (velvet curtain) border between science and art. Peale's love of both art and science led him to name his children after such people as Titian, Rembrandt and Linnaeus. See Susan Stewart, "Death and Life, in That Order, in the Works of Charles Willson Peale," in *Visual Display: Culture Beyond Appearances*, ed. Lynne Cooke and Peter Wollen (Seattle: Bay Press, 1995), 31–53. Peale's son Titian Ramsay Peale, a painter himself, was one of the first amateur photographers in the U.S. See Grace Glueck, "With an Eye for Nature and Its Exquisite Forms," *The New York Times*, 16 August 1996, C30.

of relation: species, genus, order, family, and so on.[28] The system was designed to encompass every form of life on earth: "known" and "unknown." It is a descriptive system, in which a given plant or animal is placed according to "objective" assessment – based on observable "facts" – of its similarity to and difference from previously described and placed flora or fauna. Interestingly enough, Linnean taxonomy is rooted in sex; organisms are characterized by comparisons largely based on their "male" and "female" reproductive organs.[29]

Londa Schiebinger contextualizes Linnaeus further, arguing that the search for order in nature was intimately bound up in the Enlightenment political philosophy of "natural rights," the premise on which a group of affluent white American men could base their revolutionary slogan that "All men are created equal." As Schiebinger notes, however, there was a deep reluctance among many of those calling for such rights to extend that line of reasoning very far; it seemed that "an appeal to natural rights could be countered only by proof of natural inequality."[30] The study of nature was the task of science, and scientists searched for natural laws. They sought to discern nature's order or, barring such a discovery, impose an ordering system of their own. Natural scientists participated in this political reformulation by turning their attentions to "natural" hierarchies among humans, restricting application of "natural rights" to the group in power, white European (and often, propertied) males.[31]

Linnaeus expressly included human beings in his system of nature, eventually dividing *homo sapiens* into types nominally based on geographic region – American [Indian], European, African and Asian – as well as non-place-specific wild men and monsters. While Linnaeus did not invent the concept of "race," he offered scientific structure for racial discourse, particularly since his categorical descriptions elevated the European type and denigrated the others. For example, Linnaeus distinguished *Homo sapiens afer* (the African black), which he described as "ruled by caprice," from *Homo sapiens europaeus*, "ruled by customs."[32]

---

28 As Donna Haraway notes, Linnaeus's system was "a way of ordering the relations of things through their names" (*Primate Visions*, 81). Linnaeus's own given name was Carl Linné.

29 Linnaeus's taxonomic system used the reproductive parts of plants to determine their place in the order of nature, fixing a metaphor of heterosexual gender complementarity onto plants. Linnaeus also reproduced the social hierarchy of males over females within his system; a plant's male parts (stamens) determine its class and its female parts (pistils) determine its order, with class taxonomically "higher" than order. See Londa Scheibinger, *Nature's Body: Gender in the Making of Modern Science* (Boston: Beacon Press, 1993).

30 Schiebinger, *Nature's Body*, 143. See also Nancy Stepan, *The Idea of Race in Science: Great Britain 1800–1960* (London: Macmillan, 1982).

31 As Schiebinger notes, there is nothing "natural" in Linnaeus's systemization of nature on the basis of reproductive parts. Indeed, contemporary rival naturalists, in particular Buffon, considered Linnaeus's system overly abstract and artificial; Buffon challenged the entire practice of system building. Yet Linnaeus prevailed.

32 Gould, *Mismeasure of Man*, 35.

Pratt further implicates Linnaeus's system of classifying plants and animals with the impetus for the development of scientific anti-conquest travel writing. In the wake of the acceptance of Linnaeus's systemization of nature in the second half of 1700s, naturalists traveled to the Americas, Africa, Asia and Australasia to identify and add more species to the master taxonomy. In the process they naturalized European bourgeois presence there and helped construct European authority in a global arena.

Linnaeus's system of nature achieved the acceptance and longevity that the system suggested by Comte de Buffon, his French rival, did not. Yet as Nicholas Thomas argues, it was Buffon who established the language for presenting humans in a taxonomic framework so as to "typify" people by culture.[33] Thomas credits Buffon's *Natural History* with popularizing, if not initiating, a reductionist approach to describing species. In his depictions, Buffon rendered an entire species as an individual representative: not *dogs* (in general), but *the dog*. "In virtually every case, the animal is not treated as a collectivity, but reduced to a singular standard specimen: 'The ape … is as untractable as he is extravagant. His nature, in every point, is equally stubborn.'"[34] Thus not only does a single, male individual represent the entire designated group, but the description, once it has established identifying features of physical form, is then focused on behavioral characteristics, that is, social character. The consequences of this typography can be seen in Johann Reinhold Forster's description of the people he encountered on Cook's second voyage to the Pacific (1772–75). He portrayed them with "a Buffon-like conflation, whereby the experience and perception of the knower are taken not merely to express his response to the other but [the other's] essential nature: like beauty, the 'outline of the body' is something primarily in the mind of a viewer, who takes the other person as something to be seen, rather than as an actor of the same kind as himself."[35]

The singular language of "types"

This rendering of people as "types," then, naturalizes their (apparent) behaviors. Social and physical traits, the latter identified as such according to visible physical characteristics, are condensed into a single individual, who then represents the entire group. That representative performs an abstracted role by embodying the cultural essence of the group, which is also identified with a specific geographical location much the way plants are mapped onto their native soil. Brought into the body, the social or cultural is rendered "natural"; in nineteenth- and early twentieth-century science, that embodied cohesion of culture and biology was coded as "race." As

33 Nicholas Thomas, *Colonialism's Culture: Anthropology, Travel and Government*. (Princeton: Princeton University Press, 1994).

34 Ibid., 82. Thomas quotes from George Louis le Clerc, Comte de Buffon, *A Natural History, General and Particular*, trans. William Smellie (London: Thomas Kelly, 1866), i, 531.

35 Ibid., 85.

Nancy Stepan has pointed out, "To a typologist, every individual belonged to an undying essence and bore in some way the characteristic features of this essence, however much these features were disguised." In the search for a reducible, knowable, singular substance, the nineteenth-century scientist's task, "was to explore not variation, but the stable essences behind variation."[36]

Abstraction from a visible singular subject to larger collective essence has also been an important idea in art, "the old principle of art transforming the particular into the universal."[37] Representations of "types" in art corresponded with the quest for the "ideal" long expressed by European and European-American artists.[38] Croce, for example, considered ideality "the quintessence of art."[39]

In a point not irrelevant to an examination of *National Geographic* photographs, in classical European tradition it was the nude that embodied the search for the ideal.[40] Usually a young adult, healthy and attractive (the last a varying attribute depending on the tastes of the day), nudes in painting and sculpture represented an abstract ideal. Although nudes could be claimed to represent purity and virtue, art historian Kenneth Clark has argued that "no nude, however abstract, should fail to arouse in the spectator some vestige of erotic feeling … and if it does not do so, it is bad art and false morals."[41]

National Geographic was not the first popular family-friendly monthly to feature photographs of the nearly naked. *Munsey's* feature "Modern Artists and Their Work," was sometimes illustrated with photographs of nearly nude artist's models in artistic poses.[42] The photographs strove to represent Clark's "good art," intertwining qualities of eros, beauty and "type." It was the high-art framework in which these photographs were displayed that allowed for their publication. Still, *Munsey's*, the first of the 10-cent monthlies, was not one of the genteel or "quality" magazines like *Harper's Monthly* or the *Century* after which the *National Geographic Magazine* had modeled itself.

The artistic near-nudes in *Munsey's* were white, summoning thoughts of pale marble statues. The nearly nude subjects of *National Geographic*'s photographs were most specifically not white, though sometimes they were pictured so as to resemble statues. For example, a 1924 photograph of "An Australian Aboriginal with sea hawk's eggs" shows a young loincloth-clad Australian with a "splendid physique" standing on a rock as if on a pedestal, next to a huge nest, out of which

36 Stepan, *Idea of Race in Science*, 94.

37 John Berger, "Understanding a Photograph," in *Classic Essays in Photography*, ed. A. Trachtenberg, 294.

38 Martha Banta, *Imaging American Women: Idea and Ideals in Cultural History* (New York: Columbia University Press, 1987); Graham-Brown, *Images of Women*.

39 Croce, *Guide to Aesthetics*, 15.

40 Clark, *The Nude*.

41 Ibid., 29.

42 Schneirov, *New Social Order*, 88–89. Schneirov's illustrative example is from 1894.

he has taken the eggs.[43] He is the ideal in his physique, in his brave action, in his representation of true Aboriginal life.

Donna Haraway points to the convergence of artistic and scientific ideal types in her discussion of the stuffed animals at the Museum of Natural History. Taxidermy did more than provide three-dimensional visual representation of faunal taxonomic diversity; "taxidermy was about … the unblemished type specimen," the artistic effort of making nature "true to type."[44]

For Max Weber, the scientific abstract type proved a useful device for the social sciences. By 1904, however, when he first published the essay "Objectivity in Social Science and Social Policy," Weber already found the concept "ideal type" somewhat troublesome. Trying to rescue the "ideal type" from what he called "the widely discussed concept of the 'typical' which has been discredited though misuse,"[45] he argued that the analytically useful abstract "type" should not be confused with a real, existing example of either an essential or an average "type." Generic concepts or classes, Weber argued, "are pure mental constructs, the relationships of which to the empirical reality of the immediately given is problematical in every individual case."[46] The abstract nature of an ideal type made it relational to, not prescriptive of, the concrete content on which it was based.

Weber's exposition suggests the prevalence of tangible manifestations of "types." "Typification," as Nicholas Thomas notes, "proceeded most powerfully through the evocation of the singular character."[47] Mary Louise Pratt calls this reduction of the collective to the singular a "standard apparatus" of travel writing. In the narrative equivalent of photographing individuals to represent their immutable cultural whole, the narrator constructs an abstract person as the incarnation of that whole. Pratt chose a segment of John Barrow's *Travels into the Interior of Southern Africa in the Years 1797 and 1798* to illustrate how this process of "typification" worked in travel texts. "In his disposition he [the Bushman] is lively and chearful; in his person active. His talents are far above mediocrity; and, averse to idleness, they are seldom without employment."[48] In a proto-Galtonian gesture of the imagination, Barrow has constructed a composite figure, situating the cultural personality and physiognomy in a single male actor.[49] Even Barrow's use of the collective "they" in

---

43 M.P. Greenwood Adams, "Australia's Wild Wonderland," *NGM* 45 (March 1924), 339. The photograph is by William Jackson, Nor' West Scientific Expedition of Western Australia.

44 Haraway, *Primate Visions*, 38.

45 Max Weber, *The Methodology of the Social Sciences*, trans. and ed. Edward A. Shils and Henry A. Finch (Glencoe, IL: The Free Press, 1949), 100.

46 Ibid., 103.

47 Thomas, *Colonialism's Culture*, 86.

48 Pratt, *Imperial Eyes*, 63.

49 While the composite subject is gendered male, the subject's *country* is frequently gendered female. Russia is "she"; the Russian, "he." For more on the female iconography of nation, in which iconographic representations of nations have been cast as womanly figures (such as France's Marianne) and republican citizenry confined to real male bodies, see Lynn A.

the same passage bands together a super-organic entity that acts in timeless fashion. The abstracted *he/they*, Pratt explains, "characterize anything "he" is or does not as a particular event in time, but as an instance of a pregiven custom or trait."[50]

Typification, that rendering of timeless cultural essence through the reduction of a collective into a "unitary entity which could be known," appears with almost surprising consistency from the mid-eighteenth century through today.[51] Indeed, it is not difficult to find examples of "this language of typification," in other time periods,[52] nor need such typing be reserved for people perceived as vastly different. Ralph Waldo Emerson's 1856 *English Traits* reveals this tendency:

> The Englishman speaks with all his body. His elocution is stomachic – as the American's is labial. The Englishman is very petulant and precise about his accommodation at inns, and on the roads; a quiddle about his toast and his chop, and every species of convenience, and loud and pungent in his expressions of impatience at any neglect. His vivacity betrays itself, at all points, in his manners, in his respiration, and the inarticulate noises he makes in clearing the throat; – all significant of burly strength.[53]

Texts of the *National Geographic Magazine* also reveal "this language of typification." For example, Maynard Owen Williams, the subject of the following chapter, adopted this format in his *National Geographic* article, "Latvia, Home of the Letts." While the article examined the new nation then emerging from the events of World War I and its subsequent independence from Russia, Williams shifted from the immediacy of the situation to pronouncements about the essence of the Letts. Giving up on their country, he wrote,

> is not the nature of the Letts or the Russians. They stick it through, the Lett by determination, the Russian by Philosophy. The Lett is friendly, shrewd, conservative, persevering, without the "Nichevo" spirit of Russian fatalism. Having waited so long for his opportunity, having won his freedom against such odds, he is determined to make the most of it."[54]

Williams slides from plural to singular in this cultural imagery. As Thomas notes, the condensed singular is a critical component of typing. While descriptions of people

---

Hunt, *Politics, Culture and Class in the French Revolution* (Berkeley: University of California Press, 1984); Joan B. Landes, *Women and the Public Sphere in the Age of the French Revolution* (Ithaca: Cornell University Press, 1988); Mary P. Ryan, *Women in Public: Between Banners and Ballots, 1825-1880* (Baltimore: Johns Hopkins University Press, 1990); Marina Warner, *Monuments and Maidens: The Allegory of the Female Form* (New York: Atheneum, 1985).

50 Pratt, *Imperial Eyes*, 64.

51 Thomas, *Colonialism's Culture*, 85.

52 Ibid., 89.

53 Ralph Waldo Emerson, excerpt from *English Traits*, in *The Norton Book of Travel*, ed. Paul Fussell (New York: W.W. Norton & Co., 1987), 368.

54 Maynard Owen Williams, "Latvia, Home of the Letts," *NGM* 46 (October 1932), 443.

and cultures in the plural (*they* are like this) also contribute to typing, and have also been commonly used, the condensed singular provided the conceptual precedent for photographic types.

As evidenced by school geography textbooks, the idea of cultural essentialism was nearly hegemonic in early twentieth-century geography.[55] Some fairly standard examples can be found in *Dodge's Comparative Geography*, a high-school level textbook published by Rand McNally in 1912. "The Germans are naturally home-loving and quiet, painstaking and thorough; they are less impetuous then the French." And again, in France, the "rural life of one province is often different from that of another, because the people are of different ancestry and hence of various dispositions."[56] In this last construction, biology (ancestry), place (province) and behavior (disposition) are all intricately interwoven into a cultural essence (rural life).

David Green refers to this "reduction of sociological processes to the immutability of nature" as the "biologisation of history."[57] In this process, "sociocultural differences among human populations became subsumed within the identity of the individual human body. In the attempt to trace the line of determination between the biological and the social, the body became the totemic object, and its very visibility the evident articulation of nature and culture."[58] With visibility so key in marking types, in is not surprising that photography was so quickly adopted as a tool for collecting and identifying human "types." Nicholas Thomas notes further that, "As physical forms, as characters, as bearers of customs and distinctive material artifacts, others were visible as knowable types."[59] Thus photography reinforced and codified the use of types, providing veritable physical, visible specimens to represent the embodiment of an entire complex of cultural and biological meaning.

Roland Barthes's remarks in *Camera Lucida* are pertinent here: "It is rather as if I had to read the Photographer's myths in the photograph, fraternizing with them but not quite believing in them. These myths obviously aim … at reconciling the Photograph

---

55 See Schulten, *Geographical Imagination in America*. Academic geography of the period had similar leanings, although with greater complexity and often, more environmental causality. Derek Gregory (*Geographical Imaginations*, 42), for example, discusses Vidal's work – "a contemplation of the timeless landscapes of rural France – 'a geography of permanences'" in which "'the peasant is *part* of the landscape'" – as providing the sort of theoretical basis with which to supply Dodge with his textbook proclamations.

56 Richard Elwood Dodge, *Dodge's Comparative Geography*, revised ed. (Chicago: Rand McNally, 1912), 261, 238. "In the large cities of France, and especially in Paris, the people are very polished and fond of excitement. The latter characteristic is expressed in their manner of speech, in their movements, and in their fondness for political struggles. In rural communities, on the contrary, the people are more slow of speech and action, more satisfied with the monotony of their daily life, and less easily excited." (p. 238).

57 David Green, "Classified Subjects: Photography and Anthropology: The Technology of Power," *Ten-8* 14 (1984), 32.

58 Ibid.

59 Thomas, *Colonialism's Culture*, 86.

with society (is this necessary? – Yes, indeed the Photograph is *dangerous*) by endowing it with functions, which are … to inform, to represent, to surprise, to cause to signify, to provoke desire."[60] One of the most enduring "photographer's myths" in the *National Geographic Magazine*, indeed in early twentieth-century ethnographic photography in general, and one intimately and *dangerously* aimed at "reconciling the Photograph with society" was that of human "types." Signifying the scientific quest for natural order, the political desire for social order, and the aesthetic yearning for the ideal, "types" populated the *National Geographic Magazine* from the turn of the century to the turn of World War II. While certainly not its sole genre of photography, or even ethnographic photography, "type" pictures played a significant role in *National Geographic*'s history, and provide a clear example of how photography melded science and art.

Photography's burden of truth

Questions of whether photography was more properly considered an art form or a tool of science, or to which degree it belonged to either science or art, surrounded photography from its inception in the mid-nineteenth century.[61] Susan Sontag characterizes the history of photography as "the struggle between two different imperatives: beautification, which comes from the fine arts, and truth-telling, which is measured not only by a notion of value-free truths, but by a moralized ideal of truth-telling, adapted from the nineteenth-century literary models and from the (then) new profession of independent journalism."[62] Rochelle Kolodny stretches Sontag's formulation into a more defined tripartite schema: romanticism, realism, and documentary. Romanticism dwells in the world of art, dealing with ideals and essences. Realism, in the world of scientific fact, works within a positivist, empirical ideology. Documentary corresponds to Sontag's "moralized ideal of truth-telling," situated in the world of social science and political comment.[63] The three strands, of course, are rarely isolated, instead mixing and merging in complex ways.

From the beginning, there was little question that photography gave visual representation a distinctive realism that paintings and drawings could not produce. As true-to-life, as evocative as a painting could be, it could never guarantee, in

---

60 Roland Barthes, *Camera Lucida: Reflections on Photography*, trans. Richard Howard (New York: Hill & Wang, 1981), 28.

61 See for example Alan Trachtenberg, ed., *Classic Essays on Photography* (New Haven: Leete's Island Books, 1980); Alan Thomas, *Time in a Frame: Photography and the Nineteenth-Century Mind* (New York: Schocken Books, 1977); Sontag, *On Photography*.

62 Sontag, *On Photography*, 78.

63 Rochelle Kolodny, "Towards an Anthropology of Photography: Frameworks for Analysis," (Masters Thesis, 1978, Department of Anthropology, McGill University), as discussed by Elizabeth Edwards in her introduction. Elizabeth Edwards, ed., *Anthropology and Photography 1860–1920* (New Haven: Yale University Press, 1992), 9–10.

the way a photograph did, that "the thing has been there."[64] The "photographer is bound by simple truth – happily that is an important, if not the all important principle in representation," commented an early British photography journal; the photographer "can neither add anything to adorn his picture, nor remove anything that is offensive" so that the photograph appears "as the exact transcript of nature."[65] And for nineteenth-century positivists, notes Allan Sekula, "photography doubly fulfilled the Enlightenment dream of a universal language: the universal mimetic language of the camera yielded up a higher, more cerebral truth."[66]

Many people believed that with its precision, detail and technological impassivity, photography, more so than painting or drawing, captured a person's true character.[67] The first widespread application of photography in the United States, in the mid-nineteenth century, was portraiture. Likenesses of the "great and the good" were publicly displayed in picture galleries and published in photo albums. These were individuals singled out for their fame, usually conflated with their considerable moral and/or financial worth. Photography was soon domesticated and incorporated into urban middle-class culture, with daguerreotype miniatures replacing their painted predecessors. Whether they were miniatures made for lockets or larger pictures for household display, the subjects of these pictures were intimately known to those who possessed them.[68]

Almost as soon as it was invented, the camera became the favored device for recording images of far-away people and places, becoming an integral part of Western practices of travel ranging from state- and society-funded explorations and scientific endeavors, to tourist excursions. "Prints, sketches, travellers' tales, *objets d'art* – not one of those earlier forms of evidence could match the power of witness of [photographs]."[69] One of the earliest books of photographs to be published was *Excursions daguerriennes: Vues et monuments les plus remarquables du globe*, in 1841, just four years after Daguerre invented his photographic process.[70] Photographs from abroad became immediately popular. They were published variously as illustrations in travel books, as collections in books, or individually as cartes-de-visite, postcard-sized portraits that were precursors to tourist postcards, or

64 Barthes, *Camera Lucida*, 76.

65 S. Bourne, "On Some Requisites Necessary for the Production of the Good Photograph," *Photographic News* 3 (1859), 308. Quoted in Elizabeth Edwards, "Photographic 'Types': the Pursuit of Method," *Visual Anthropology* 3 (1990), 237.

66 Allan Sekula, "The Body and the Archive," in *The Contest of Meaning: Critical Histories of Photography*, ed. Richard Bolton (Cambridge: MIT Press, 1989), 352.

67 Trachtenberg, *Reading American Photographs*.

68 Ibid.

69 Thomas, *Time in a Frame*, 23.

70 It contained engravings based on daguerreotypes. Du Camp's 1852 *Égypte, Nubie, Palestine et Syrie*, featured calotypes commissioned by the French government. Sontag, *On Photography*; Banta and Hinsley, *From Site to Sight*, 39.

as landscape "views."[71] By 1860, "scarcely an expedition left for a frontier region without a cameraman or two."[72]

Photography developed deeply within the political, martial and economic processes of Western imperialism. As a medium that presumed to capture reality, photography was seen as a valuable tool by those who sought to assess the qualities of the hitherto unknown or unfamiliar people encountered in the course of empire-building. Colonial administrators wanted to be able to identify and control their subject peoples. Scientists wanted to build their knowledge of humanity, and of racial categories in particular. Citizens of the bourgeois metropole wanted to tweak their imaginations with images of the strange and exotic. Often, the same photograph served multiple demands. "Ethnographic studies made by government photographers went both to museum archives and to bookshops on fashionable boulevards."[73] Imperial governments provided access to scientists; scientists' theories of race and culture supported imperialist pursuits.

Dehumanizing visual human data

The advent of photography came just as the expeditionary sciences of geography and anthropology were legitimizing and institutionalizing themselves in new ways. With their exclusive focus on people, anthropologists grew interested in the possibilities of photographic portraiture. But they were not interested in the famous or in the familiar and known; they sought photographs of "others," unfamiliar by personal acquaintance or by achievement's reputation. They were seeking visible statements of a larger "racial" character in the "other," not photographs of individual characters. As E.B. Tylor commented in 1876, "the science of anthropology owes not a little to the art of photography."[74]

To nineteenth-century practitioners of the "science of man," or anthropology, photography produced a valuable new form of data.[75] Anthropologists, evolutionary biologists and other scientists concerned with humans turned to photography to help them classify human racial "types." "Photography came to play a central and complicitous role in what became the dominant concern of late nineteenth-century anthropology, the articulation of race and racial differences."[76] Popular and scientific

71 Thomas, *Time in a Frame*. Some cartes-de-visite were sold in stores; those were usually photographs of celebrities, but photos of unknown exotic peoples were also commercially marketed.

72 Ibid., 30.

73 Ibid., 21–22.

74 Quoted in Roslyn Poignant, "Surveying the Field of View: the Making of the RAI Photographic Collection," in Edwards, ed., *Anthropology and Photography*, 55, from E.B. Tylor, "Dammann's Race-Portraits," *Nature* (6 January 1876), 184–5.

75 Edwards, "Photographic 'Types'"; Green, "Classified Subjects"; Banta and Hinsley, *From Site to Sight*, 46.

76 Green, "Classified Subjects," 31.

beliefs in racial fixity were remarkably persistent despite the obvious centrality of change in evolutionary theory and the recognition among scientists that racial groups as they perceived them were hybrids of varying histories.[77] At a time when "scientists of man" were avidly measuring human intelligence by filling skulls with birdseed – again, to determine racial, and sex, differences[78] – scientists deemed "objective" and "realistic" photographic recordings useful tools for comparison, especially if done systematically. British scientists Thomas Henry Huxley and John H. Lamprey, for example, each devised systems of photographic measurement in which the naked subjects were posed in a series of positions either before a grid (Lamprey) or with a ruler (Huxley).[79] Francis Galton developed composite photography – what he termed "pictorial statistics"[80] – in his effort to discover the ultimate example of the "type," whether criminal, consumptive, or Jew.

Studying the use of photography in nineteenth-century criminology, Allan Sekula identified the two major formations in the history of photographic realism as the channeling of images into emblematic types, and the filing system for unique images.[81] Searching for "the criminal type," late nineteenth-century social scientists such as Cesare Lombroso, Havelock Ellis and Francis Galton, poured over photographs of people arrested on charges ranging from loitering to murder, seeking common traits and often enough, finding them. Because the "criminal type" was considered organically distinct from the normal and non-criminal, the criminologists presumed that physiological characteristics common to criminals would be revealed through this method of compiling and comparing photographs. What is of interest to Sekula is the system of photograph management, the way in which the information was organized by a "bureaucratic-clerical-statistical system of 'intelligence'": the filing cabinet.[82] A kind of photographic equivalent of Dewey's bibliographic system for libraries, this method, developed by Alphonse Bertillon of the Paris Prefecture of Police, relied on a taxonomic translation of anthropometrics based on eleven bodily

---

77 Edwards, "Photographic 'Types'"; Stepan, *Idea of Race in Science*; George W. Stocking, Jr., *Race, Culture and Evolution: Essays in the History of Anthropology* (New York: Free Press, 1968).

78 Gould, *Mismeasure of Man*.

79 Green, "Classified Subjects"; Edwards, "Photographic 'Types'"; Christopher Pinney, "Classification and Fantasy in the Photographic Construction of Caste and Tribe," *Visual Anthropology* 3 (1990): 259–287; Frank Spencer, "Some Notes on the Attempt to Apply Photography to Anthropometry During the Second Half of the Nineteenth Century," in Edwards, ed. *Anthropology and Photography,* 99-107.

80 Sekula, "The Body and the Archive," 368.

81 Ibid., 373. Sekula's twin photographic approaches complement the "two commensurate acts" that Trachtenberg (*Reading American Photographs,* 9) finds are required "to make the order of things visible: ... clarity of representation, and formal placement of the represented thing within a strictly defined spatial form." Both modes of photographic semiotics resonate loudly in the *Geographic*'s portrayal of ethnic "types" differentiated from one another by their organic affiliation to unique places.

82 Ibid., 351.

measurements.[83] Bertillon's police photo archive was "the first rigorous system of archival cataloguing and retrieval of photographs," adapted and emulated in one form or other by "a bewildering range of empirical disciplines, ranging from art history to military intelligence" between 1880 and 1910.[84]

"The surveillance of the gaze was one of the chief instruments of domination, whether of the criminal, the insane, or the subject peoples of the Empire," notes David Green.[85] As many others do, he situates photography within the development of the surveillant state theorized by Michel Foucault. Within the North American context, governmental geological and boundary surveys in the Western regions netted thousands of photographs. Topographical "views" showed beautiful wilderness, with no visible human presence. And yet, the same photographers, for example, W. H. Jackson, official photographer for the American Geological Survey of Territories, filled formal photograph albums with portraits of the Native Americans they had rendered invisible in the landscape.[86] The relationship between "ethnographic" photography and colonial control is explicitly demonstrated in the making of the 8-volume *The Peoples of India* (1868–75). Originally an informal project suggested by Lord Canning, the British Governor-General of India, army officers were encouraged to photograph various Indian "types" for souvenirs. But the 1857 Sepoy uprising brought the project into the official terrain of the Political and Secret Department. Individual subjects form the majority of the images. They are identified by costume, occupational objects, weaponry and caption as a "typical" specimen of the described sort. The descriptions then proceed to assess the "character" of the "type" especially as it related to their governability by the colonial power.[87]

---

83 Ibid. Sometimes a "criminal" functioned as an individual, rather than as a type. Hence, the mug shot. Developed by Bertillon, the mug shot consisted of a frontal and a profile photograph of the apprehended person's face, taken within precise settings of light and camera distance. The method aimed to catch repeat offenders and the photographs were designed to identify persons who might otherwise get away by using a variety of disguises and fake names.

84 Ibid., 373. According to Sekula, "the archive became the dominant institutional basis for photographic meaning … roughly between 1880 and 1910" extending as well into the homes of the middle-class. Popular stereoscopic cards, for example, presented pictorial groupings based on themes, often geographic, rather than in any linear narrative. "There were always more images to be acquired, obtainable at a price, from a relentlessly expanding, globally dispersed picture-gathering agency" (p. 374). This form of photographic archive may have achieved its zenith in *National Geographic*. Not only did the magazine assiduously collect and print photographs from as many places in the world as possible, most of the magazine issues themselves presented a random selection of photographed places. In addition, *National Geographic* has functioned in popular culture as a multi-volume encyclopedia of sorts, routinely collected and compulsively stored in people's personal archives.

85 Green, "Classified Subjects," 32

86 Poignant, "Surveying the Field of View," 55.

87 Pinney, "Classification and Fantasy"; Poignant, "Surveying the Field of View."

The slippage in the use of "type" photographs between scientific, governmental, military, and popular underscores the importance of discourse in maintaining and reproducing the colonial/imperial/carceral state that developed in Europe and the "neo-Europes" in the modern period. While the purposes to which they were put were ostensibly different, the photographs shared a language of making the strange-and-unfamiliar the strange-and-familiar. They established distance by creating a one-sided visual proximity, feeding into a framework of naturalized hierarchy among human groups. Although some anthropologists, ethnologists and geographers devised meticulous procedures for photographic human measurement, others were satisfied with the formal portraits taken by scientists, studio photographers and travelers. Louis Agassiz, for example, turned the hand-tinted cartes-de-visite "of native types" sold to tourists into data for his analysis of racial types.[88] As James Ryan notes, "The broad currency of the 'type' across discourses of both art and science meant that photographs could signify as 'types' in a whole range of ways," including accuracy in comparisons of measured human difference and as signs of character.[89] According to Alan Thomas, "such photographs of native people, treated as curios, even wonderful, subjects to be gathered into a comprehensive visual taxonomy, reflect the European base of photography and the general association of the camera, a precision instrument, with scientific purposes."[90]

Organizations concerned with science and with the "diffusion of knowledge" helped to popularize "scientific" photography of "native people." The British Association for the Advancement of Science recommended the use of photography in its *Notes and Queries on Anthropology*, first published in 1874 "for the use of travellers and residents in uncivilized lands," and later editions spelled out particular methods in which to pose and photograph the subjects.[91] Even those subjects spared the ritual of scientifically posing naked had their photographed image read and studied in Western cosmopolitan centers as "types." E.B. Tylor was instrumental in establishing the use of photography for members of Britain's Royal Geographical Society. In his essay for the Society's publication *Hints for Travellers*, which remained in print for over 50 years, Tylor essentially argued that since people of a simple-living tribe all look alike anyway, unlike "the individualized faces of a party of Europeans," the photographer, "if he has something of the artist's faculty of judging form, may select groups for photography which will fairly represent the type of a whole tribe or

88 Banta and Hinsley, *From Site to Sight*, 46.

89 Ryan, *Picturing Empire*, 151.

90 Thomas, *Time in a Frame*, 32.

91 Green, "Classified Subjects," 34. Green quotes from the 1899 (third) edition of *Notes and Queries*, published by the Royal Anthropological Institute: "With regards to portraits, a certain number of types should always be taken as large as possible, full face and square side view; the lens should be on a level with the face, and the eyes of the subject looking straight from the head should be fixed on a point at their own height from the ground … When the whole nude figure is photographed, front, side and back views should be taken; … It is desireable to have a soft, fine-grained, neutral-tinted screen to be used as a background. This screen should be sufficiently light in color to contrast well with the yellow and brown skins."

Figure 3.1 "Types in La Paz," ran the caption from February 1909
The photograph was taken during Harriet Chalmers Adams's inaugural trip to South America
(Harriet Chalmers Adams/*National Geographic* Image Collection)

nation."[92] "'Types' were very seldom named or identified beyond the very general; tribe, place of origin, or trade, for example, – 'A Burmese Beauty', 'A Typical Native', 'A Native Warrior.'" As Edwards notes, "The stress was on the generality as represented in one specimen. Photographically, the 'type' is expressed in a way which isolates, suppressing context and thus individuality… . Through photography the specimens, 'types', were neutralized and objectified for scientific use to be interpreted and reinterpreted."[93]

---

92 Quoted in Ryan, *Picturing Empire*, 148.

93 Edwards, "Photographic 'Types,'" 241. The "scientific" use of photography to delineate types found a new application in the mid-twentieth century with W.H. Sheldon's vast archive of nude "posture photos" of first-year college students at elite universities. Sheldon's collection of photographed bodies was connected to his formulation of the three main body types: ectomorph, mesomorph and endomorph, and their corresponding personality types. (Ron Rosenbaum, "The Great Ivy League Nude Posture Photo Scandal," *The New York Times Magazine*, 15 January 1995, 26–31, 40, 46, 55–56.)

Creating *National Geographic*'s archive of ethnographic photography

In the early decades of the twentieth century, The *National Geographic Magazine* often identified the local inhabitants of particular places as "types." Functioning as cultural synecdoche – a part standing in for the whole – individual people were stripped of their personal identity and made to don the mantle instead of their perceived cultural/ethnic/ racial identity. The entire world was divided into "types": a Hungarian type, a Negrito type, a Zulu type. On a page of the July 1907 *National Geographic*, the reader would find a photo of a "Typical Eskimo Dog," and on the following page, a "Typical Face of Eskimo Woman."[94] Fourteen years later, readers were introduced to "A Korean Type" and "A Peasant Type of Anatolia."[95]

People did not have to be young, female, naked, dark-skinned, or even foreign to be fitted comfortably into types. A 1921 article on the lives and environment of New England fishermen repeatedly expresses admiration for the "splendid inherited qualities of the type" that enable them to do this life-threatening work generation after generation.[96] The photographs depict white fishermen aboard their ships. They appear strong and stern as they look at the horizon. The rendering of this localized occupation group as a natural "type" is built on a perceived distance of class between the fishermen and the presumed white-collar middle and upper class *National Geographic Magazine* readership. Although the stereotype is a laudatory one, its flattery is patronizing.[97]

There are several ways in which photographs of "types" appeared in the *National Geographic Magazine*. Editors might select photographs already displaying "types," or they could manipulate other photographs to render them as "types." In the latter approach, editors could frame any portrait as a "type" through the photograph's caption; label a photographed person a "type" and he or she *becomes* a "type."[98] Another framing device that editors used was the cropping of a photograph. Cropping provided a focus on a single person at the expense of extraneous context.[99]

---

94 *NGM* 18 (July 1907), 462, 464.

95 *NGM* 46 (October 1924), 371, 459.

96 Frederick William Wallace, "Life on the Grand Banks: An Account of the Sailor-Fishermen Who Harvest the Shoal Waters of North America's Eastern Coasts," *NGM* 40 (July 1921), 7.

97 The *Geographic*'s presentation of fishermen in New England as "types" is not at all unusual for the time period. Geographer J. Russell Smith's 1925 college textbook North America, for example, mentions such "types" as "hardy yankees, slothful southerners, independent yeomen, and dim-witted immigrants." James Curtis, *Mind's Eye, Mind's Truth: FSA Photography Reconsidered* (Philadelphia: Temple University Press. 1989), 8.

98 W.J.T. Mitchell has elucidated the importance of the caption to the photograph in his discussion of Edward Said and Jean Mohr's 1986 photo essay on Palestinians, *After the Last Sky*. At a U.N. conference, they were allowed to hang the photographs, but prohibited from providing any captions. "Context, narrative, historical circumstances, identities and places were repressed … " Mitchell, *Picture Theory*, 315.

99 While I have no evidence that *National Geographic* refocused a photograph in such a manner, there is evidence that the magazine's illustrations and picture editors have

Of the "type" photographs, many were produced by professional photographers, government and military agencies and ethnographic organizations, and purchased by or loaned to the *Geographic*. "Type" photographs were also taken by *Geographic* contributors expressly with the readers and the editors of the magazine in mind.

Photographs presented to or purchased by the magazine as types could come from government sources, such as the War Department of the United States, which oversaw U.S. administration of the Philippines. The *National Geographic Magazine*'s summary of the Philippine census and its reprint of many of the original report's photographs, discussed in the previous chapter, is one such example. The magazine also purchased "type" photographs from professional photographers, many of whom had studios located in the colonized places featured in *National Geographic Magazine* articles. Such photographs were often picked up by contributors on their travels, sometimes to illustrate their own story, sometimes just to add to the Geographic's vast photo archive. Maynard Owen Williams, for example, "bought a few photographs of Greek types from Zogrophos" in 1928.[100] Years later, on a trip to Hawaii, Williams met "master Kodachromist" William F. Sullivan. Writing to then associate editor John Oliver La Gorce, Williams reported that he had "given [Sullivan] a card to Frank [Fisher, *Geographic* pictures editor] which he will send in with a few of his shots. He can make them any way we want them but is selling scads of them as Kodaslides at 75¢ each – surf-riders, hula gals, flowers, beaches, volcanoes."[101]

One classic "hula gal" appeared in a 1924 article on Hawaii by editor Gilbert H. Grosvenor. The portrait of a "Pure-blooded Hawaiian girl wearing the costume of past generations," is of a seated young woman of about 13 years wearing a grass skirt and nothing else – though she holds a ukulele.[102] A photograph similar in its formal studio rendition of the "beautiful young Pacific Island woman type" appears in a 1919 *Geographic*. Sitting demurely in front of a blurred background, the topless

---

doctored photographs. Over the years, they have played with skin tone (Tom Buckley, "With the *National Geographic* on Its Endless, Cloudless Voyage," *New York Times Magazine*, 6 September 1970, 10–11, 13–14, 18–21), covered up genitalia (Bryan, 100 Years), and erased urine markings (Haraway, *Primate Visions*). John Raeburn has revealed the power of editorial cropping in *Look* magazine's version of the Photo League's "Harlem Document," in which the meaning of photographs (by a Harlem resident of Harlem residents) was substantially altered – and not in the interest of the subjects – by the way in which the magazine cropped the original photographs. (John Raeburn, "African-Americans Mediated: The Photo League, the "Harlem Document," *Native Son*, and *Look* magazine," paper presented at the American Studies Association annual meeting, Kansas City, Missouri, 31 October – 3 November 1996.)

100 MOW, 24 February 1928, Line-a-Day Diary, Williams Collection.

101 MOW to John Oliver La Gorce (hereafter JOL), 26 September 1941, Williams Collection.

102 Gilbert H. Grosvenor, "The Hawaiian Islands – America's Strongest Outpost of Defense – the Volcanic and Floral Wonderland of the World," NGM 45 (February 1924), 125.

Figure 3.2 "A daughter of a dying [Marquesan] race – beautiful, luxuriant hair, fine eyes, perfect teeth, a slender, graceful form, a skin of velvet texture and unblemished figure."

(From *National Geographic Magazine* October 1919. L. Gauthier/*National Geographic* Image Collection)

young woman holds a bouquet of flowers in her hand; a long necklace of shells is carefully draped on the outside of her breasts. The woman is, the caption tells us, "A daughter of a dying [Marquesan] race – Beautiful, luxuriant hair, fine eyes, perfect teeth, a slender, graceful form, a skin of velvet texture and unblemished figure."[103] Instead of personalizing the subject of the photograph by providing her name or a biographical snippet, the caption underscores and overdetermines the

103 John W. Church, "A Vanishing People of the South Seas – the Tragic Fate of the Marquesan Cannibals, Noted for Their Warlike Courage and Physical Beauty," *NGM* 36 (October 1919), 276.

position of the photographed subject as rare, and dehumanized, object. The form of these photographs is reminiscent of the late nineteenth- and early twentieth-century "studio fantasies," often manufactured as postcards, that featured "exotic" women. As Malek Alloula has noted, these photographic versions of European "Orientalist" painting drew on European male erotic fantasies of the harem, providing instant access to a collection of beautiful young "dark" women richly adorned with jewelry and cloth, or in a languorous state of near undress.[104] Indeed, some of the studio portraits printed in the *National Geographic Magazine* were *precisely* the same images that were marketed for the use of French Foreign Legion soldiers and other Europeans seeking souvenirs of their stay in Algeria. Linda Steet points out that a non-credited photograph in the January 1914 issue of the *National Geographic Magazine*, captioned "Two Ouled Naïls in Characteristic Garb" and followed by an instructive paragraph on the economic and social position of dancing girls in Algeria, is in fact a French colonial postcard. Other postcard models appear in the pages of the *National Geographic Magazine* as well.[105]

As Steet argues, studio photographs were only the most obvious of the *Geographic*'s construction of ethnographic "realities." As such, they offer conspicuous testimony against *National Geographic*'s claim to objective naturalism in its photographs. Many such photographs appeared in the magazine uncredited, especially in the years before World War I. For example, a model in a photo in the March 1906 issue captioned "A Moorish Belle," also appeared in a postcard of the period.[106] The credit line for the *Geographic* photo reads "Photo from David G. Fairchild." Since most of the credited photographs in the magazine specified that they were taken *by* the photographer, Fairchild apparently procured it. The actual photographer, like the model herself, remains, at least to the reader, anonymous.[107]

---

104 Alloula, *The Colonial Harem*; de Groot, "'Sex' and 'Race'"; Graham-Brown, *Images of Women*. Mitchell, *Picture Theory*, 309, notes that in Alloula's book, "'Aestheticization,' far from being an antidote to the pornographic, is seen as an extension of it, a continuing cover-up of evil under the sign of beauty and rarity."

105 See Steet, *Veils and Daggers*, 37–42.

106 Ibid., 38–39.

107 NGM 17 (March 1906), 137. The photograph was probably purchased on Fairchild's 1901 trip to North Africa as a USDA botanical researcher. Fairchild was a photographer himself, however, and took his own pictures of fascinating locals, in particular the "enormously fat Jewesses" of Tunisia. "On my return to Washington I showed my photographs of these enormous Jewesses to a young man by the name of Gilbert Grosvenor whom I met at the Cosmos Club. He was building up a little magazine as the organ of the National Geographic Society and his flair for the spectacular induced him to publish these pictures." Fairchild, *The World Was My Garden*, (New York: Scribner's Sons, 1938), 195. Through Grosvenor, Fairchild met and married Marian "Daisy" Bell, Elsie Bell Grosvenor's younger sister.

In the years prior to World War II, *National Geographic* amassed thousands of photographs from sources including story authors, government files, and by 1919, a few paid staff photographers. The magazine also used submissions by "freelance" photographer-adventurers, as well as photographs purchased by Geographic staffers on their travels. By the 1920s, photographers on assignment were making hundreds of photographs for each story and writing captions for each picture. While *National Geographic* articles were liberally illustrated, only a small portion of the photographer's yield was actually published. "A man who takes 1,000 pictures a month and gets three printed, begins to take the caption matter lightly," grumbled a staff photographer in 1927.[108] The photographs that made it to publication were carefully selected by the editorial staff, in particular, the illustrations editor.

Although the *Geographic* prided itself on being up-to-date, that concern was largely reserved for the Society's maps. Most maps, illustrating specific articles, were published within the magazine, but larger fold-out maps were regularly included in a separate folder that came with the magazine.[109] A different standard was applied to photographs. That this was the case reveals the *Geographic*'s modus operandi in which politics were seen to be in constant flux but culture was viewed as an embodied constant. The latest maps informed readers of the most recent arrangements of political borders, the changing names of places for political – often, in this period, colonial – reasons, and the places newly "discovered" by explorers from the Western metropoles. But old photographs were used to maintain an aesthetic of timelessness. If culture and race were essential qualities, then what difference would it make if a 20-year-old photograph were used to illustrate an ethnographic "fact"? Thus the belief in the timelessness of the mother-child relationship can be seen in the use of two different photographs of Japanese women with children. One photograph, of a woman pushing a baby carriage, had been published as early as 1895 as part of a stereoscope card but ran in a 1932 *Geographic.*[110] Another photograph, showing a woman with two babies on her back, was taken by Eliza Scidmore and appeared in both 1907 and 1917.[111]

Grosvenor's 1924 article on Hawaii featured several photographs from Henry W. Henshaw, a U.S. Biological Survey ornithologist who had lived in Hawaii from 1894 to 1904.[112] One photograph, featuring a grass house, noted in its caption that it was taken in 1890 and that the only known such house "to be found in the islands

---

108 MOW to Franklin L. Fisher, 14 May 1927, NGS Records Library.

109 For more on National Geographic maps, see Schulten, *Geographical Imagination in America*.

110 Judith Babbitts, "Half-Tones and Half Truths: One Hundred Years of American Photographs of Japan," paper presented to the American Studies Association annual meeting, Pittsburgh, 9-12 November 1995.

111 Schulten, *Geographical Imagination in America*, 161–162.

112 Henshaw was part of the social circle of federal scientists that had dominated the Geographicin its early years. See Edward William Nelson, "Henry Wetherbee Henshaw, Naturalist 1850–1930," *The Auk: A Quarterly Journal of Ornithology* 49 (October 1932): 399–427. Available at elibrary.unm.edu/sora/Auk/v049n04/p0399-p0427.pdf, accessed 3 September 2006.

to-day is carefully preserved in the Bishop Museum."[113] The passage of time is also referred to in the portrait of a young bare-breasted woman with the caption "Pure-blooded Hawaiian girl wearing the costume of past generations,"[114] though not in the minimalist caption "Spearing turtle, Hawaii" for a photograph of a man, seen from the back, dressed only in a loincloth and aiming a spear at the water. The use of these photographs suggests that the *National Geographic* was willing to reach back into time to provide readers with pictures of near-naked natives even as it advertised that it showed readers "a true reflection of the customs of the times."[115] The representational efforts of *National Geographic* bought into – and sold back – what Marianna Turgovnick calls the West's "temporal illusion," based on "an ethnographic sense of existing primitive societies as outside linear time."[116]

The denial of change implicit in the use of old photographs and stock images as contemporary illustrations specifically demarcated the idealized, stable, and ultimately knowable "them" from the continually progressing "us."[117] In the case of Japan, for example, idealized portrayals of Japanese peasants coexisted with the anti-Japanese immigration laws and social activity in the United States. Admired and appreciated by Americans as long as they were in their stable and proper place "over there," the very same people were seen as threatening harbingers of instability once they crossed the ocean to the United States. Reframed as "immigrant," they became no longer admired, but despised. Moreover, the consistent portrayals of Japan as rural and "premodern" worked to mask that country's significant rate of urbanization, industrialization and military expansion.[118]

The *Geographic* collected photographs from governments, ethnographic organizations (such as museums), studio photographers and other secondary sources, but its regular contributors also provided the magazine with photographs of "types." Often a random person on the street could stand as a representative "type," but contributors sometimes went further in their efforts. Maynard Owen Williams, for example, rationalized his own machinations as a photographer by appealing to artistic principles in his support of notions of cultural essence and timelessness. "I have been guilty of posing pictures, of bringing old costumes forth into the light, of adding something of MOW to them," Williams wrote to illustrations editor Franklin Fisher in 1927. "But those are pardonable defects. The people I posed can speak better that way. The costumes I found can suggest a more

113 Grosvenor, "The Hawaiian Islands," 120.

114 Ibid., 125.

115 Ibid., 39.

116 Marianna Torgovnick, *Gone Primitive: Savage Intellects, Modern Lives* (Chicago: University of Chicago Press, 1990), 46.

117 Babbitts, "Half-Tones and Half Truths."

118 Ibid.

colorful day, the MOW that is added is the necessary person who had to carry the camera and pull the trigger."[119]

Williams's argument for dressing his subjects in what he took to be traditional clothing betrays the consciousness with which contributors and editors saw people they photographed as representative of or stand-ins for their designated and predetermined culture. "The people I posed can speak better that way," he wrote. But speak of what? As Steve Cagan notes, photographs are still and silent; you can see the people in photographs, but you cannot hear them speak. Photographs have worked as silencers;if any voice is heard, it is that of the photographer.[120]

Williams himself was explicit about his agency in the process. The voice that Williams seemed to believe he was enhancing was not the individual words or speech of the subject, but a supra-personal utterance, an essential sound emerging from the central core of the subject's cultural being. For Williams, "modern" clothing served only to disguise this cultural essence. In a 1926 letter to Frank Fisher bemoaning change in Palestine, it appeared that for Williams, a defining characteristic of Palestinian women was that they wore a particular locally made scarf; once they started wearing imported scarves, they became less Palestinian.[121] For Williams, being Palestinian (or Malaysian, or Cretan) was both a cultural and place-centered condition. As people in a place became more obviously globalized, their unique local qualities diminished. More troubling for the photographer, as Williams might have put it, was the lack of a visual identifying marker.

Indeed, the whole exercise of exploration and travel is dependent on distinctiveness, things that set apart one place from another, one group of people from another. That distinctiveness has rested, ultimately, on visual comparison. The *Geographic*'s preference for "traditional" tableaux suggest its desire to make clear and perpetuate the difference between the "advanced," modern world for which it was a representative and the rest of the world, forever less developed, always of another time. In addition, many of the magazine's photographs turned social subjects into picturesque objects, a process that worked to evaporate the political context in which the subject was embedded.

Much effort was put into showing the "natural" and immutable cultural essence of the *National Geographic Magazine*'s ethnographic subjects, and the timelessness of place-bound and racialized culture. Williams found it necessary to coax his subjects into wearing clothes their parents or grandparents may have worn in order to show the *Geographic* readership the "true" substance of their overseas contemporaries. The *Geographic*'s editors reached into their archives to reprint photographs decades old. Interestingly, the current *National Geographic*'s Flashback feature, referred to earlier, replicates the effect of rendering "exotic" cultures unchanging. The 1919-era photograph stands decontextualized. There are no contemporary photographs with

119 MOW to Franklin Fisher, 14 May 1927, NGS Records Library.

120 Steve Cagan, "Photography's Contribution to the 'Western' Vision of the Colonized 'Other,'" paper presented at the Center for Historical Analysis, Rutgers University, 1990.

121 MOW to Franklin Fisher, 19 March 1926, NGS Records Library.

which to compare it, and despite its stated existence as an early twentieth-century photograph, it is textually framed as a scene that an eighteenth-century naturalist might have actually witnessed. It is therefore unclear whether it is the content of the photograph that makes it anachronistic, or the photograph's style: black and white, carefully posed, picturesque and strikingly erotic.

"Clouds of fantasy and pellets of information"

"Cameras did not simply make it possible to apprehend more by seeing," Sontag noted. "They changed seeing itself, by fostering the idea of seeing for seeing's sake."[122] Photography became an important device of the "world-as-exhibition" that emerged full-blown in Europe and North America by the second half of the nineteenth century. It was a world "ordered up ... before an audience as an object on display, to be viewed, experienced and investigated."[123] Steeped in late nineteenth-century modes of photographic representation of the world's people, the *National Geographic Magazine* carried these images and the ideologies on which they were based into the thick of the twentieth century.

Formal "types" photographs, frequent in the *National Geographic Magazine* through the 1930s, declined around the time of the Second World War. Far fewer captions labeled the subject as a "type" in the 1940s, for example, than they did in the 1920s. This change most likely reflected a move in the sciences away from racial typology, a change spurred in part by the discovery of the gene and the recognition of the horrors of Nazi "eugenic" practice.[124] Another factor in the turn away from formal "type" photographs may have been the development of new techniques in photography; lighter cameras and faster shutter speeds enabled photographers to take more active, less formal pictures.[125] But while there were fewer formal "types" photographs, *National Geographic* continued to favor portraits of people "in their native dress." Lutz and Collins have shown in their study of *National Geographic*'s photography after 1950 that post-World-War-II *National Geographic* did not drift far from its established modes of "soft" and "hard" primitivism in its portrayal of "Third World" peoples.[126]

Photography gave a new and heightened emphasis to visual representation in science. The technology of the camera not only enabled the physical process of imprinting on paper the subject before the lens, but gave the resulting visual image

---

122 Sontag, *On Photography,* 84.

123 Timothy Mitchell, *Colonizing Egypt* (Cambridge: Cambridge University Press, 1988), 6.

124 Stepan, *Idea of Race in Science,* 139. "Only in the 1920s and 1930s did racial typology begin to get worn away at the edges through a series of scientific, social and political changes."

125 Priit Juho Vesilind, *The Development of Color Photography at National Geographic*, M.A. thesis, Syracuse University, 1977.

126 Lutz and Collins, *Reading National Geographic*, esp. 136.

itself the stamp of indisputable reality, of *fact*. But visual representation had long belonged to the realm of art, with connotations and practices that embraced notions of the subjectivity of the practitioner, flights of fancy, and sentiment, ideas inimical to the Enlightenment objective ideal in science. Photography allowed these two separate strains of objectivity and subjectivity, of science and art, to intertwine as never before. As Sontag has argued, photographs "trade simultaneously on the prestige of art and the magic of the real."[127] Conversely, of course, photographs count on the prestige of the real and the magic of art. "They are clouds of fantasy and pellets of information,"[128] and that is precisely the formula that the *Geographic* worked so well to its advantage. By 1908, more than half of the magazine's pages were devoted to photographs.[129] In embracing the photograph as its dominant "language," *National Geographic* developed an effective and seductive strategy of innocence.

127 Sontag, *On Photography*, 63.
128 Ibid.
129 Bryan, *100 Years*, 121.

Chapter 4

Maynard Owen Williams: Contradictions of a 'Seeing-Man'

If the tables were turned and someone were looking for a representative figure to embody the physiognomic and cultural essence of "National Geographic land" in the early-to-mid-twentieth century, Maynard Owen Williams would be a prime candidate. White, Anglo-Saxon, Protestant, big and athletic, well-educated and bespectacled, friendly and affable, Williams wrote and/or contributed photographs to over 90 articles for the magazine – enough to fill five specially bound volumes. His writing tended toward the corny and his photographs aimed for the picturesque. "'Mr. Geographic" to a generation of its readers,"[1] Williams worked for the Geographic from 1919 until his retirement in 1960, spending 30 years as the magazine's chief of foreign staff. His work for the *National Geographic Magazine* included such high-profile stories as the revelation of the contents of King Tutankhamen's tomb in 1923, an arctic expedition with Donald MacMillan in 1925, and the 1931–32 Citroën-Haardt Trans-Asia Expedition, a grandiose attempt at a motorcar journey through mountainous central Asia. Williams was an immensely prolific private writer as well, sometimes writing as many as five or six letters a day.

Williams, who graduated from Kalamazoo College in Michigan, joined the small Geographic staff after serving a stint as a Protestant missionary teacher in Beirut and Hangchow, and as a foreign correspondent for the *Christian Herald* in Asia and Europe. He spent a few months as a relief worker among the Armenians in Van in the winter of 1917–1918, and incorporated some of his experiences in his second article for the *Geographic*.[2] Williams met his wife, Daisy Woods, in China, where she was also a missionary teacher. A Mount Holyoke graduate, she often moved with Williams and their eventual five children when he was stationed intermittently in Europe, the Middle East, and Washington D.C., but Williams spent a good deal of time away from home, traveling and photographing. Williams was happiest when he was "in the field,"

1 Bryan, *100 Years,* 179.

2 Williams, "Between Massacres in Van," *NGM* 36 (August 1919): 181–184. The article, published during the politicized climate surrounding the First World War, mentioned "Kurdish fiends" and "Turkish hordes" but focused on the brave spirit of young Armenian children who smile as they trudge barefoot through the snow. Despite the present tense used throughout, in a subtle but devastating final paragraph, readers discover that Williams has been describing a period prior to a Turkish massacre of Armenians. He returned in 1926 and prepared text and photos for an article, but it did not run. See http://www.nationalgeographic.pt/revista/0304/online_extra.asp (accessed 21 July 2006).

and tended to be miserable when he had to spend any great length of time in the Washington headquarters of the National Geographic Society.

Fiercely loyal to the Geographic while chronically at odds with his editors, Williams provides a prime example of the contradictions of a self-conscious American bringing his embedded vision to bear on other peoples in other places. A study of Williams's diaries and letters, read with and against his published articles indicates how "the reality of the world" was selected with great purpose for the waiting members (and other readers) at home. Initially enthusiastic about showing this world to a wide audience, Williams developed a growing ambivalence over his participation as professional "seeing-man" in making the world safe for tourism.

Williams as "seeing-man"?

"Seeing-man" is Mary Louise Pratt's name for the leading practitioner of anti-conquest narratives. The gender is intentionally male, and the term specifically implies a European male subject. It is "seeing-man" who provides Pratt with the title of her book as well, for it is "he whose imperial eyes passively look out and possess."[3] As opposed to the narrator of tales of survival or physical conquest, "seeing-man" relates his story in terms of passive vision. As in the expression "look but don't touch," looking is an innocent act, removed from obvious activity, aggression or interaction. This strategy of innocence, however, is simultaneously offset by what has been recognized as the possessive intent of European vision.[4] That which could be seen could be known and consequently coveted and claimed.

The sense of possession comes through in Pratt's discussion of what she calls the "monarch-of-all-I-survey" genre, a particularly Victorian form of seeing-man anti-conquest writing.[5] Pratt chooses Richard Burton's tale of African exploration as a prime example of the genre. Burton presented his discovery of Lake Tanganyika in a painterly and heavily aestheticized narrative centered upon the passive act of seeing; "within the text's own terms the aesthetic *pleasure* of the sight single-handedly constitutes the

3 Pratt, *Imperial Eyes*, 7.

4 See, for example, Greenblatt, *Marvelous Possessions;* Pratt, *Imperial Eyes*; John Berger, *Ways of Seeing* (London and Harmondsworth, British Broadcasting Corporation and Penguin Books, 1972); Denis Cosgrove, *Social Formation and Symbolic Landscape* (London: Croom Helm, 1984).

5 The phrase "monarch of all I survey" comes from a 1782 poem by William Cowper written from the imagined perspective of Andrew Selkirk. Selkirk had been a castaway for several years on an island over 400 miles off the coast of Chile and his popular account of his adventures, published in 1713, is believed to have been the inspiration for Daniel Defoe's *Robinson Crusoe*. Although Selkirk's unusual circumstances made him essentially the leader of a colony of one, Cowper's phrase captured what was perhaps already a recognizable link between power and visualization.

See http://www.bbc.co.uk/history/scottishhistory/europe/oddities_europe.shtml (accessed 16 July 2006) and for Cowper's poem, http://www.bartleby.com/106/160.html (accessed 16 July 2006).

Figure 4.1 Seeing-man: Maynard Owen Williams in the rigging of the Arctic-bound *Bowdoin*, 20 June 1925
(Jacob Gayer/*National Geographic* Image Collection)

value and the significance of the journey."[6] The narrative became a verbal landscape painting. For Pratt, the painting analogy plays up both the aestheticizing of the landscape and the power of perspective: "If the scene is a painting, then Burton is both the viewer there to judge and appreciate it, and the verbal painter who produces it for others… . it also follows that what Burton sees is all there is."[7] The painting analogy evokes as well John Berger's discussion of the development of landscape painting as a response to landowners' desire to encapsulate and reproduce the image of the lands they possessed.[8]

---

6 Pratt, *Imperial Eyes*, 204.
7 Ibid., 204–205.
8 Berger, *Ways of Seeing.*

Figure 4.2 "My feeling is for the beautiful and very often the beautiful as suggested by an arch of lightly leaved bamboo bent so it frames a cloud," Williams wrote to his wife Daisy
This photograph graced a 1935 National Geographic article on the Italian Riviera.
(Maynard Owen Williams/*National Geographic* Image Collection)

As with Burton, Maynard Owen Williams's pleasure of discovery lay in the act of seeing. Writing to Grosvenor in 1929 about his experience as a *National Geographic* photo-journalist, Williams noted that "the greatest 'kick' a field man can have is to carry a million and a quarter members up onto a high mountain, show them the world and say, 'It's yours, in a way it could not be without me.'"[9] But Williams inhabited a different world from Burton: the American twentieth century. A photographer as well as a writer, Williams was "both the viewer there to judge and appreciate it, and the … painter who produces it for others." Indeed, it was his job to reproduce his vision through the use of photography. Williams's joy lay not in his own individual visualization, but in his being the visual conduit for millions. Williams was conscious of his audience; he carried them with him where he went, and he went, on assignment, to seek the sights his audience wanted to see. He was *employed* to see for others, and yet his visions were regulated and mediated by his employer. Williams was a loyal *Geographic* staffer, and strongly believed in what he deemed to be the mission of the

9 Byran, *100 Years*, 183.

National Geographic Society and its magazine. Continually pushing aside his own recognition of the magazine as glossy travel literature, he framed the Geographic as a vehicle for world brotherhood, promoting "good will" and "true understanding."[10] He traveled to gather images of the world – immediately visual photographs, adjective-laden copy – that were aimed for parlors and living-rooms of middle-class America. For both personal and professional reasons, he sought the aesthetic in landscapes and in the people who were always potential subjects for his camera. His letters and diaries reveal his work to be explicitly guided by his ideas of beauty, his interpretation of the *Geographic*'s audience, and an ideology of "world brotherhood" based on his strong but churchless Christian theology. "My feeling is for the beautiful and very often the beautiful as suggested by an arch of lightly leaved bamboo bent so it frames a cloud," Williams wrote to his wife Daisy in 1937.

> I waste endless films trying to express a mood – not mine, but nature's. After that I return to the Office, to listen to telephones, to have the stuff dreams are made of measured as merchandise, 750 words to a page and three pages of pictures to one of type. The amazing thing is that so much of the world's beauty and interest survives such methods.[11]

While Williams generally embraced the "innocence" within his production of embedded narrative, at times he not only recognized but worried about his role in the assertion of Western hegemony. Throughout his tenure at the magazine, Williams periodically ran up against differences with his editors regarding the *Geographic*'s approach and even had occasion, as we shall see later, to question the model of innocence which he had taken to be the *Geographic*'s as well as his own.

'Mr. Geographic': the right man for the job

To a great extent, though, Williams embraced his role as visual conduit of information about places and people in distant settings and was deeply loyal to the *National Geographic* project. He supported the *Geographic*'s apolitical pretensions and humanist pronouncements, agreeably supplied the magazine with photographs of smiling women – with or without upper-body clothing – and worked to frame the world in aesthetic terms.

In making the move from the missionary milieu of the *Christian Herald* to the environs of science and journalism – with their attendant claims to objectivity – Williams had to re-orient his prose style to avoid the sorts of political and religious analyses that were permissible in an advocacy setting. As noted in Chapter 2, Gilbert

10 MOW to Edwin L. "Bud" Wisherd, April 1945, Item XLII, Box 10, Williams Collection.

11 MOW to DWW, 30 August 1937, Item XX, Box 3, Williams Collection. All following references to letters dated 1937, unless otherwise noted, are from letters Williams sent to his wife while he was on the National Geographic Society-Smithsonian Institution East Indies Expedition, and are in this container.

H. Grosvenor had decided early on that the *National Geographic Magazine* would express itself in a nonpartisan and objective manner, and devised seven guiding principles for the magazine, almost half of which were directed at maintaining a nearly self-righteous apolitical stance. Thus principle three, "Everything printed in the Magazine must have permanent value," was related to principle five, "Nothing of a partisan or controversial nature is printed," in that a nation's politics were deemed to be both transitory and separate from enduring culture. Principle six built on principle five: "Only what is of a kindly nature is printed about any country or people, everything unpleasant or unduly critical being avoided."[12]

This declared disinterestedness in political affairs, bathed in the rationale of "Victorian courtesy,"[13] is one way in which National Geographic asserted a studied innocence. Williams earnestly followed Grosvenor's prescription, agreeing with his editor that "we are objective in viewpoint and helpful in tone, not toward any specific faction, but toward an understanding of world life, irrespective of political questions."[14] In a later letter to Grosvenor, Williams reiterated the understanding that "'the captains and the kings depart' and politics does not fill libraries with bound volumes of The Geographic."[15]

For Williams, this commitment to presenting peoples and places without politics was consistent with his philosophy of classic humanism: that true and fair knowledge of others is the key to fostering tolerance, and that tolerance and respect for others is what the world needs to make it a better place. His humanism had a religious dimension as well. Bought up in the Methodist Christian church, he saw in faith a guide toward peace and goodness and had a profound respect for "God's world." Williams was earnest in his expressed belief that underneath the colorful trappings of specific culture, people were essentially the same all over the world. When people asked him how he got along "with all those strange people," he said, he would reply, "Show me the land where babies cry when they're happy and laugh when they are sad and I will show you a land where people are strange."[16] In notes for a possible autobiography, Williams wrote: "Adventure, to me, has not been danger, or hardship, but a revealing that men everywhere like what I like, want what I want, and realize that if we are to have it, we must give ourselves the same chance of friendship that we do in distrust."[17]

In the *National Geographic Magazine*, which "by restricting its articles to what is kindly … has made friends the world round," Williams saw his ideals of world understanding put into action.[18] The magazine served as a medium for Williams' commitment to world friendship; the theme appeared in his published articles as

---

12 Bryan, *100 Years*, 90.

13 Ibid.

14 MOW to GHG, 7 March 1927, NGS Records Library

15 MOW to GHG, Item XLII, Box 10, Williams Collection.

16 Autobiography notes, envelope v, Item XXVII, Box 7, Williams Collection.

17 Envelope v, Item XXVII, Box 7, Williams Collection.

18 MOW MS for GHG biography, 2nd version, 1960 or 1961, 12, Williams Collection.

Figure 4.3 Maynard Owen Williams believed that "by restricting its articles to what is kindly," *National Geographic* "made friends the world round"

In this previously unpublished photograph, Williams shakes hands with "a little devil-worshipping Yezidi, near Pamp, Armenia."

(Maynard Owen Williams/*National Geographic* Image Collection)

well as in his personal notes and letters. In a 1932 article about the Citroën-Haardt Trans-Asia Expedition, of which he was the National Geographic representative, he told the story of having the expedition's doctor attend to a sick girl in Xinjiang (Sinkiang at the time). Recounting the incident, he incorporated notes he took at the time: "Whether she lives or not, the time will come when all will forget this 'foreign devil' who juggled with death under the dusty trees in a night-black courtyard. But the tradition of mutual friendship may live." After nearly 60 pages of article, more than half of it photographs, Williams concluded with the striking statement that "Our aim was not to court adventures, but friendships."[19] Williams's stress on friendship over adventure is particularly interesting considering that the point of the expedition – sponsored in large measure by the automobile manufacturer Citroën – was to be the first to travel across Asia by motored vehicles. A convoy of mostly Europeans riding in motorized vehicles and mostly Kashmiri guides and servants riding on pack animals seems an unlikely method in which to cultivate friendship in Asian areas without paved roads, readily available gasoline, or much money.

Williams' emphasis on friendship was his own particular articulation of *National Geographic*'s universalist humanism, one of the magazine's most enduring strategies of innocence. As Lutz and Collins discuss for the post-World-War-II era, *National Geographic*'s humanism elicited the demonstration of positive universal human traits: familial unity, mother-child bonds, pride in labor, physical attractiveness of youth. While on one hand, such a projection of essential humanity allows for the dismantling of ideas of social hierarchy, it also belies the formation and effect of power relations. Noting that this form of humanism is not confined to *National Geographic*, but has been prevalent among museums and photographic exhibitions, Lutz and Collins cite Roland Barthes's critique of the 1955 *Family of Man* exhibition (and book) of photographs that was much like a collection of *National Geographic* illustrations. "That work is an age-old fact does not in the least prevent it from remaining a perfectly historical fact," Barthes wrote. Speaking of the "very differences in its inevitability" based on the historical development of colonial and class relations, Barthes argued that "we know very well that work is 'natural' just as long as it is 'profitable.' … It is entirely this historified work that we should be told about, instead of an eternal esthetics of laborious gestures."[20]

Like his editors, though, Williams disliked ugly "historical facts." In the same article on the Citroën-Haardt expedition, Williams told of how he "felt a bit ashamed" at photographing opium-smokers and dealers in the streets of Liangchow, "and it was a relief to turn to a jolly cobbler, whose smiling face would show friendliness even in reproduction." But Williams was, in turn, reprimanded by an English-speaking Chinese man for photographing such a humble character. A "conscientious photographer," as he refers to himself, Williams defended his subject selection, reasoning to his admonisher – and to his readers:

> This man is a worker. He is clean. He seems reasonably happy and honest. He represents the real China. He is neither imposing exorbitant taxes nor gratifying personal ambition at the expense of the poor. Far from discrediting your country, his picture will help offset stories of maladministration and hatred of "foreign devils." He and I cooperated in that

---

19 Williams, "From the Mediterranean to the Yellow Sea by Motor," *NGM* 62 (November 1932), 516, 580.

20 Lutz and Collins, *Reading National Geographic*, 60–61; Roland Barthes, *Mythologies*, trans. A. Lavers (New York: Hill and Wang, 1972), 102.

Figure 4.4 "A Cobbler of Liangchow," subject of a debate over photographic representations
Williams explained to a passerby and then to readers, that he and the cobbler "cooperated in that picture through a tacit understanding. In this limited sense, we like each other."
(Maynard Owen Williams/*National Geographic* Image Collection)

> picture through a tacit understanding. In this limited sense, we like each other. No such evidence of mutual friendliness can dishonor your land.[21]

Williams gets the last word in this exchange with his righteous pronouncement and printed photograph.

21 Williams, "Mediterranean to the Yellow Sea," 560–61.

In this epistle, Williams displays several hallmarks of *National Geographic* representation, asserted through his own particular vision. The first line of Williams's defense is that the subject is "a worker." As such, he belongs to the Geographic's global pantheon of disaggregated workers, which are indeed, as Barthes suggests, aestheticized.[22] His goodness derives in large measure from his humbleness, as Williams clarifies further; he knows his place and does not demand more than his small share. True to *National Geographic* guidelines, the subject is set up to personify only positive virtues, such as cleanliness, honesty and friendliness, and is established as a personification of a place. What does Williams actually know about this man? Really only that he is, at least at the moment that Williams sees him, fixing shoes. The rest is conjecture built out of desire. Compared with the opium dealers, certainly, this man is just what Williams wants to see: a good, friendly, ordinary Chinese man. Unlike the historically developed opium dealers, defined by a bad fad, the humble cobbler represents – to Williams, and through Williams to the readers – a timeless essence, interchangeable not only with thousands of other Chinese cobblers, but with thousands of years of Chinese cobblers. His presumed ordinariness and unhistoricized identity as a worker are what allow him to represent the "real China." (Williams's concerns with essential verity are discussed in greater depth below; the "cooperation" and "tacit understanding" between the photographer and his subject reappear further still.)

Williams's particular emphasis on friendship was his own articulation of *National Geographic*'s general global humanism. In correspondence with Gilbert Grosvenor, it is Williams, and not the editor-in-chief, who refers to the *Geographic*'s mission as one of international friendship. What was nominal promotional policy for the magazine was creed for Williams. Williams took special pleasure in the idea of a global membership of the National Geographic Society. He delighted in finding a National Geographic member overseas, and mentioned such incidents in his letters to his editors as well as in his own diaries.

For the most part, Williams was an optimist, willing to give people the benefit of the doubt. When he found himself having negative thoughts about a place, such as its "spoiling" from tourism or modernization, or the difficulty he may have had communicating with its inhabitants, he would step back and remind himself how lucky he was to be doing what he did for a living. Catching himself complaining in a letter to his wife, he remarks that his "ill-tempered gossip" – complaints about the social habits of his colleagues – seem to suggest "that this is not – as it quite possibly is – the best of all possible worlds."[23] The way to deal with all the problems in the world, he counseled, was to find one thing that bothers you most and work to change it, make it better, or get rid of it.[24] Williams's "one thing," the problem to which he dedicated his life to changing, was what he saw as enmity based on

22 See also Rothenberg, "Voyeurs of Imperialism."

23 MOW to DWW, 20 March 1937, Williams Collection.

24 Envelope e, Item XLIII, Box 10, Williams Collection.

cultural misunderstanding. Reflecting on his life and work upon his retirement in 1953, Williams wrote:

> I feel that this good old world should know itself better. I feel that many of our fears are groundless. I feel that God knew his business in putting a little variety into the human race. I feel that, during a long life of travel and writing, I have tried to inspire understanding and trust rather than fear and hate.[25]

The artist is a missionary

An especially important universal characteristic for Williams was that of beauty. Williams appreciated the role and value of photography for "truth-telling," but for him the "truth" of his subject came not through any inherent objectivity in the use of the camera but through the "beauty" in what the camera managed to capture. Williams recognized the *National Geographic Magazine*'s genre as travel writing, as well as its persona as popular science; in his letters, he distinguishes it from both journalism and what he called "highbrow science."[26] As far as his own photography was concerned, however, Williams saw himself as an artist, or perhaps more modestly, a would-be artist. "The photographer is an artist," he wrote to his wife in 1937, "in that he not only pushed the shutter release and lets God's sun and man's skill record the scene but that he catches that scene in a significant mood or moment whose significance may be apparent only to himself."[27]

Williams expressed a growing sense of frustration with his ability to capture in photographs the "beauty" that he saw before him. "A few of my pictures, if seen by sensitive folk, will conjure up visions of sea and cloud and jungle somewhat comparable to the beauty that was here when my camera and I looked on it," he wrote to Daisy. "But what a wastage [?] between the thing itself and the poor reproduction of it! … I, Mr. X, saw such and such a scene and felt thus and so about it. With such skill and sincerity as I possess I tried to share its beauty with others. And this ugly travesty of the thing is the result. Beauty, passing through my eye and hand and being, became this puerile parody of grandeur and majesty."[28]

"Perhaps," he wrote, "it is that the artist is – willy nilly – a missionary. Knowing beauty, he feels others should know it too."[29] A former child evangelist and later a teacher in overseas missionary schools,[30] Williams's enthusiasm for the [Methodist] Church had soured somewhat in the wake of what he considered its un-Christian

---

25 MOW to Kalamazoo College President Smith, 4 February 1953, Williams Collection.

26 MOW to JOL , 7 March 1923, NGS Records Society.

27 MOW to DWW, 13 November 1937, Williams Collection.

28 MOW to DWW, 23 August 1937, Williams Collection.

29 MOW to DWW, 23 August 1937, Williams Collection.

30 Williams made reference to his childhood zealotry, recalling a scene in church when he was 13 and calling for his fellow parishioners to "Come to Christ." MOW to DWW, 2 April 1937, Williams Collection.

support for killing in the First World War. Although a regular church-goer after the war, he remained wary of theological authority. But his religious training ran deep, and Williams carried his spiritual impetus into his work, interpreting his mission as a writer-photographer for the *National Geographic Magazine* as one of capturing the beauty of what he called "this good old world." Williams aligned himself with the artist, who, "sensing the glory of God as shown in line and color and atmosphere, … tries to convert money-grubbers from their apathy and deadness. And, having felt the surge of something beautiful through his spirit he knows his inability to transmit or translate it to others. But he can't help trying."[31]

For Williams, the truth of a photograph lay in the larger truths to which he subscribed. He followed the dictum "beauty is truth; truth, beauty." Far from taking a scientific approach to photography, perceiving the camera as objective recorder, he instead used the camera as a device in which to frame small samples of the world's beauty: mountain landscapes, humble villages, smiling young women. Williams pressed for the artist's path to truth: "For Truth is not an animal to be raped but a spirit to be wooed."[32] Both literal and psychological sunniness were key ingredients in Williams's *National Geographic Magazine* photographs, shedding light on the world, highlighting the beauty in the visible scenes before him and his camera. "A great deal of one's sense of beauty," he commented to his wife, "is a sense of good cheer and good spirits."[33]

As a photographer, Williams sought scenes and people he deemed most aesthetically satisfying, positioning himself, his camera, and even his subjects in such a way so as to enhance their "natural beauty." Framing a landscape through a tree's branches, or by waiting for the sun's light to hit at just the right angle, or by having villagers don their seldom-worn traditional clothing was only to bring out the deeper truth of the subject. "I have been guilty of posing pictures, of bringing old costumes forth into the light, of adding something of MOW to them," Williams wrote to illustrations editor Franklin Fisher in 1927. "But those are pardonable defects. The people I posed can speak better that way. The costumes I found can suggest a more colorful day, the MOW that is added is the necessary person who had to carry the camera and pull the trigger."[34]

Williams's argument for dressing his subjects in what he understood to be traditional clothing reveals the consciousness with which contributors and editors saw people they photographed as the representatives of or stand-ins for their designated and predetermined culture. "The people I posed can speak better that way," he writes. But speak of what? The voice that Williams appears to believe he is enhancing is

---

31 MOW to DWW, 23 August 1937, Williams Collection.

32 MOW to DWW, 30 September 1937, Williams Collection. Williams's characterization of the aggressiveness of Science in pursuit of Truth echoes Francis Bacon's comment that Science must take nature "by the forelock," quoted in Carolyn Merchant, *The Death of Nature: Women, Ecology and the Scientific Revolution* (New York: Harper & Row, 1980), 170.

33 MOW to DWW, 30 September 1937, Williams Collection.

34 MOW to Franklin Fisher, 14 May 1927, NGS Records Library.

Figure 4.5 "Island, Plain and Mountain Furnished These Costumes for Athenian Maids"
To enhance visual variety and reveal receded cultural traditions, Williams and other *Geographic* photographers sometimes turned to theaters and dance troupes.
(Maynard Owen Williams/*National Geographic* Image Collection)

not the individual words or speech of the subject, but a supra-personal utterance, an essential resonance exuding from the central core of the subject's cultural being. Here the tendency to render photographed subjects ahistorical is not merely an idealistic aestheticization, but an overt erasure of historical change. Williams felt that by coaxing his subjects into wearing clothes their parents or grandparents may have worn, he could more effectively show the *Geographic* readership the "true" substance of their overseas contemporaries. In so doing, of course, he was also clarifying his subjects' difference from the magazine's American readers, continuing constructions of the "other." As a photographer, Williams helped to create the image of cultural essence and timelessness that the *National Geographic Magazine* used as one of its most prominent and effective strategies of innocence.

Williams pulled back somewhat from his historical manipulations as his career at the Geographic progressed. His letters home in 1937, when he was the *National Geographic* correspondent for the National Zoo's animal-collecting expedition in the Dutch East Indies, record little of the largely sartorial orchestrations mentioned in his 1920s diaries and letters. Indeed, Williams expressed indignation over "faking" photographs by dressing subjects "in dress to which they are not accustomed," asserting that he had "long fought against faking or 'doctoring' change of a subject … . it is a passion with me." An irritated Williams said of the expedition's leader, National Zoo director William Mann, that his "frequently repeated disdain for the intellectual puerility of photographers doesn't get my goat. But his assertion that a photographer can't understand the iniquity of faking did." Mann had pointed out as fake pictures in *Natural History* magazine by Martin Johnson, who with his wife and partner Osa, was probably the best-known wildlife movie-maker in the early twentieth century. For the *Natural History* photographs, Johnson had "dressed up his boys in dress to which they are not accustomed and gave them blow-pipes which they don't know how to use." Williams noted with derision that despite Mann's umbrage at Johnson's staged photographs, he "is planning to have Liau[?] Gadi, our Borneo Dyak hunter, lay aside his gold spectacles and wear 'native' dress when he arrives in America."[35]

Yet despite taking Mann to task for his staging of timeless otherness, Williams continued to do the same. On the same 1937 expedition, Williams arranged, often with payment, for villagers to perform full-costume dances for the benefit of his camera and the *National Geographic* readership. Although color photography was not yet able to capture swift movement, dance offered the vibrancy of activity to black and white photographs. Under the more laborious process of color photography, posed dancers displayed their special costumes that were often very detailed and elaborate. As Lutz and Collins note, even in the second half of the twentieth century, *National Geographic* used dance to emphasize ritual in its representation of peoples and places, since ritual is considered one of the prominent representations of a cultural essence.[36]

Though later in life Williams refrained from active efforts to dress his subjects "in clothing that they did not usually wear," he continued to seek out landscapes and people who represented an enduring premodern "essence." By Williams's aesthetic standards, recent innovations in architecture or clothing had two strikes against them; they were not "picturesque," and they muddied a clearly defined reading of the place or culture.

> In Ramallah, imported fringed shawls are supplanting the lovely embroidered scarfs which were the most beautiful in Palestine. The automobile is raising the devil with the "Biblical museum" aspect of Palestine. The Jewish immigrants are changing styles in the

---

35 MOW to DWW, 3 August 1937, Williams Collection. Martin and Osa Johnson made several film-making (and hunting) safaris, mostly in Africa, including one with Kodak founder (and expert marksman) George Eastman.

36 Lutz and Collins, *Reading National Geographic*. This representation of cultural essence, though, is not necessarily for show to outsiders. Ritual performance and the costumes attending ritual are representations of cultural continuity for the people of that culture, whether captured for display by outsiders or not, and in the twentieth century could as well be read as resistance to globalization.

whole region. Now, from the photographer's standpoint, all this is bad. It is a case of now or never … these are precious days for photographers of Palestine.[37]

But the process of change itself seems to have been as tenacious as any cultural "essence" Williams attributed to his beloved "Bible lands." More than 25 years after despairing over the end of an era, Williams once more bemoaned the photographer's fate. "The news in Asia is a story of swift and epochal change," Williams commented in the 1950s. "But the picture-lover still prefers pictures of Good Shepards, patient donkeys, supercilious camels and veiled women – scenes of Biblical charm to him but anathema to new governments whose backwardness has been emphasized by friend and foe… . to the traveler, as such, no modern apartment house has the charm of the teeming market-place or the lush delta where blue-black water buffalo wallow in muddy pools."[38]

In his fourth decade of working for the *National Geographic Magazine*, Williams was still seeking out subjects who spoke to him of a constant past. Testifying to the strength of long-held images, Williams mused that "A long gowned peasant girl, carrying a burden on her head toward a palm-fringed oasis, has a poetic touch that is remembered long after the modern buildings are forgotten."[39] In Williams's aesthetic world, the image which evokes the past – and which is threatened with extinction – is that which actually lives on, while the pulsing modern subject which threatens to erase the precious image of the past, is ephemeral, demolished by the mind's eye of the beholder. The predetermined image, that expected – and desired – by photographer and reader alike, is the one that continues to blot out the unexpected and undesired.

Williams was thinking of his subjects in terms of their aesthetic value early in his tenure at the *Geographic*. "Corsica is infinitely beautiful, as wasteful of adjectives as a Packard Twin Six is of gas, but the people are superficially lacking in interest," he complained to Grosvenor in 1923. "I have seen just one woman here who was even good-looking, and as far as costumes, widow's weeds top the list as far as attractiveness is concerned. Never before have I visited a country where the land surpassed the people in interest, and it constitutes a new problem for my camera, as well as my typewriter."[40]

That Williams was directing his plaint to his editor suggests that his interest in good-looking women and colorful costumes was in some measure professional.

---

37 MOW to Franklin Fisher, 19 March 1926, NGS Records Library.

38 Envelope v, Item XXVII, C, Box 7, Williams Collection.

39 Ibid. Williams may have been a by-the-books pastoralist, but some of his contemporaries found inspiration in the buildings of modernity Williams found so antithetical to poetry, such as Carl Sandburg's 1916 *Chicago*: "Stormy, husky, brawling,/City of the Big Shoulders/ … I turn once more to those who sneer at this my city, and I give them back the sneer and say to them:/Come and show me another city with lifted head so proud to be alive and coarse and strong and cunning." Carl Sandburg, "Chicago," in Alexander W. Allison, et al., ed., *The Norton Anthology of Poetry*, revised ed., (New York: W.W. Norton, 1975), 963.

40 MOW to GHG, 11 January 1923, NGS Records Library.

From both the photographs and his notes, a major goal of Williams's photographic pursuits seems to have been, as he might have said, "feminine charm."[41] Notes from his 1926 trip to Palestine and later diary entries – only a few of many – indicate the extent of his pursuits:

> 30 December 1925: Arrived at Tartus in early afternoon and spent rest of day driving Moslem ladies to cover.
>
> 25 February 1926: Unsuccessful attempt to find pretty women at Mary's Well but one girl runs into a strange house thinking we intend to abduct her… . Visited village near Jenim where women are not photographed.
>
> 1 March 1926: Out to Ramallah Mrs. Kelsey helps me get some pictures but several women fail her. In desperation we go to Girls' School and stumble on beauty.
>
> 27 October 1929: Drama to Serres with stop at Pentapolis where time was largely wasted trying to get photos of girls wearing cut-velvet aprons and strings of gold coins of many vintages. Driver would get a whole bazaar aroused and of course girls were shy.
>
> 17 November 1929: Sunday chase for Corfu costumes. At Pelecka, three mediocre looking women pose for money and others refuse to pose. At Kynopiastes I induce a few women to pose and others follow. At Gastouri we get pictures of a recent bride in wedding finery evidently hired for the occasion.

Bali's breasts without shame

Williams found photographing women in Bali, in 1937, a much easier task, and to his own sensibilities at least, more rewarding. "I am shameless enough to admit that the firm young breasts of Bali give it much of its extraordinary appeal," he wrote to his wife.[42] The breasts, although Williams seems to have missed the connection in his excitement, belonged to young Balinese women. Indeed, Williams became so taken with the brown breasts of Bali that they were the focus of several letters home, prompting his wife to suggest that he write stories for the *Geographic* on "Breasts and Religion" and "Wild Breasts I Have Known."[43] In another letter to Daisy, Williams mused:

---

41 MOW to JOL, 23 December 1941, Item XLII, Box 10, Williams Collection. Williams was commenting upon Luis Marden's photographs of Panama in the November issue: "his color shots show that marriage hasn't spoiled his ever-ranging eye for feminine charm."

42 MOW to DWW, 23 August 1937, Williams Collection.

43 MOW to DWW, November (date unclear), 1937, Williams Collection. Indeed, in letters home Williams conducted a lengthy evaluation of the status of women in some of the world's major religions, writing about how the exposure or concealment of women's breasts was symbolic of a culture; free breasts represented free women, while the more hidden a women's body, the fewer the freedoms and social privileges she could enjoy. "The breasts of Bali are a delight because they go with the open air, the free stride, the happy frank look of women who are beings, not slaves, – of men or of convention." [MOW to DWW, 23 August

> Last night I looked over some films made by a very clever photographer, back from Bali. And went to bed thinking of firm, round, satiny brown breasts. I suppose Mary Woolley would call such thoughts "erotic." And why not? … priggish folks use the word "erotic" as though their parents had never felt the quiver of Eros… . These smooth round breasts are blessed with shamelessness, not because they are wanton but because they are without shame. It was pleasant to think on them, even in bed.[44]

It is interesting to see that here, too, and perhaps more obviously, Williams employs a "strategy of innocence." While he expounds to his wife on his erotic nighttime fantasies about other women's breasts, he not only naturalizes his feelings (eros is natural; eros is ordinary), but frames the subjects themselves as of a pure nature. They are "without shame" because they are without guilt, and Williams will not be made to feel guilty for his innocent reflections on breasts exposed innocently. "That I enjoy seeing pretty breasts is beside the point," he insisted.[45]

Williams' equation of his erotic feelings with innocence relies heavily on conventional associations of nature and "people of nature." People living in tropical places in particular had long been described in European and neo-European travel literature as "children of nature," a trope based on both romantic and evolutionary notions and which was used comfortably in the *National Geographic Magazine* for many years.[46] The often lush agricultural or forest landscapes and the bare clothing of the people who lived there contributed as well to frequent Edenic references. In either rendition, tropical peoples were framed as literally innocent of civilization, without the "sin" of advanced knowledge, and thus without the shame of nakedness that came only after the fall of Adam and Eve.[47] Williams brings in the rhetoric of

---

1937, Williams Collection.] Given the greater joy and spirit of women in bare-breasted societies, one could then argue that they also made for better, livelier photographic subjects on the merits of their happiness alone. Daisy Williams most likely took title inspiration from Ernest Thompson Seton's popular 1898 book, *Wild Animals I Have Known*.

44 MOW to DWW, 26 May 1937, Williams Collection.

45 MOW to DWW, 23 August 1937, Williams Collection.

46 See Rothenberg, "Voyeurs of Imperialism." Crosby uses "neo-Europe" to designate the non-European places settled extensively by Europeans: the Americas, Australia, and New Zealand. See Alfred W. Crosby, *Ecological Imperialism: The Biological Expansion of Europe, 900–1900* (New York: Cambridge University Press, 1986).

47 The equation of nakedness and innocence in references to "natives" – particularly women – of tropical contact zones has a long history. See, for example, Kolodny, *The Lay of the Land*; Smith, *European Vision*; Tiffany and Adams, *The Wild Women*. A 1774 poem by John Courtenay, inspired by European men's tales of Tahitian women, expresses this common theme:

> Naked and smiling, every nymph we see,
> Like Eve unapron'd, 'ere she *robb'd the tree*
> Immodest works are spoke without offense,
> And want of decency shews innocence
> (from Smith, *European Vision*, 32).

edenic innocence by writing to his wife that, "As yet, only the fig-leaf sarong is demanded by modesty in this paradise."[48]

Oddly enough, few of Williams's photographs of bare-bosomed women made it into the series of articles drawn from his 1937 travels. A number of photographs of bare-breasted women illustrated the articles, but most of them are credited to people other than Williams. While Williams has a photograph of a busy public fountain-bath, the adult Muslim women shown bathing are mostly robed; it is a woman, Lillian Schoedler, who gets credited for the coveted photograph of nude women and girls at a public bath, in Bali.[49] And photographer Burton Holmes contributes a full-page finale of smiling young women in "the usual everyday dress – a skirt held at the waist by a bright-colored sash, because her legs must be concealed, and a headcloth."[50] Perhaps the most conspicuous addition to Williams's article on Bali is a photograph credited to Screen Traveler; although the caption would have one believe that the bangled baby is the focus of the picture, the young mother's full breast – and little else of her beyond a part of her shadowed serene-Madonna face – is at least as much the centerpiece.[51]

In addition to employing the strategies of innocence of science and sentiment to legitimize *National Geographic's* publication of photographs of bare-breasted women (see previous chapter), the magazine relied on familiar notions of race. That such a mainstream publication as the *National Geographic Magazine* could display such photographs at all had much to do with racial distinctions, prominently based on skin color. For an article on Tahiti, for example, *National Geographic* editors doctored a photograph to "darken down" a topless Tahitian women they judged too fair-skinned for appropriate publication.[52] Significantly, it is not "Balinese breasts" that Williams waxes over, but "brown breasts." "Bali and Java, though off my beat, gave me a lot of good shots, especially nice brown skin in Kodachrome," he wrote to his wife. "The skin I'd have loved to touch – and didn't."[53]

Williams aestheticized race, and skin color in particular. In one letter, Williams speaks glowingly of a beautiful woman – half Dutch, half Bandanese – of whom he hopes to take a color photograph. "For monochrome wouldn't do her justice. Why

---

48 MOW to DWW, 26 May 1937, Williams Collection. Daisy's response to her husband's fetish is left largely up to conjecture, since while she saved all of his letters, the whereabouts of her letters to him are not in the Williams Collection. Williams frequently ended such letters with declarations of love or longing for her, often saying that he was waiting to get back into her arms again, remarking on her own lovely "milk-white" or "ivory" skin tone.

49 Williams, "Netherlands Indies: Patchwork of Peoples," *NGM* 73 (June 1938), 678; Williams, "Bali and Points East: Crowded, Happy Isles of the Flores Sea Blend Rice Terraces, Dance Festivals and Amazing Music in Their Pattern of Living," *NGM* 75 (March 1939), 314.

50 Williams, "Bali and Points East," 352.

51 Ibid.

52 Buckley, "Endless, Cloudless Voyage."

53 MOW to DWW, 23 November 1937, Williams Collection.

won't the world accept brown skin at its proper value?"[54] What may have been in some measure a call for racial harmony comes across more as a plea for Western recognition of the aesthetic pleasures of dark skin. Although his letters reveal a respect for the many brown-skinned people whom he met on his travels and took umbrage at American prejudice against Asian immigrants, Williams the artist-photographer clearly valued what he saw as exotic beauty – and found the exotic erotic.

On the whole, in line with his earnest humanism, Williams's racial attitudes were among the more progressive of those emanating from the pages of the *National Geographic Magazine*. Williams tended to complain about individuals rather than condemn an entire group, and had as many critiques of the behavior of the colonizer as that of the colonized. As devoted as he was to world understanding, however, Williams also slipped into familiar patterns of racial and racist thought. Although he more frequently had complimentary things to say about the native inhabitants he observed and worked with – including those he hired to serve as translator, driver, or baggage-carrier – he sometimes groused about their laziness or "silly" behavior. "Lazy, conceited children these Malays are," he wrote to his wife about Malay adults, annoyed at his difficulty in getting the photographs he wanted.[55]

What is most striking in Williams's notes is a subtle but pervasive negativity associated with American blacks – not Africans (Williams's African travel was limited to Mediterranean North Africa), but African-Americans. In pre-*National Geographic* articles, Williams used stereotypes of African-American culture as referents in describing the more foreign cultures of which he wrote. Using a brand of ethnic humor often dismissed by those not its target as "gentle fun," Williams used these stereotypes to denigrate the subject of his observations, comparing the many Filipino men who "dress in becoming taste" to the others who "wear neckties that would turn a 'down South darkey' green with envy."[56] A "peaceful and dirty little village" in Russia with a hard-to-pronounce, consonant-filled name is compared to "a six pound colored baby being called George Washington Abraham Lincoln Woodrow Wilson Jackson."[57]

African-Americans are, for Williams, the familiar exotic. They stand as the standard model of the exotic other, physically and culturally distinct from white European-Americans, as well as geographically situated in distinct regions such as the American South or New York City's Harlem. But they are literally too close to home to be truly exotic, and for Williams, as for most of his *National Geographic* colleagues, this supposed familiarity apparently permits contempt. Again, Williams's derisive comments regarding blacks are framed largely in aesthetic terms. Calling the people in one area "too unpicturesque to photograph," Williams explains that the

---

54 MOW to DWW, 25 September 1937, Williams Collection.

55 MOW to DWW, 23 September 1937, Williams Collection.

56 Williams, "The Philippine Carnival," circa 1917, 483, envelope e, Item XXVI, Box 5, Williams Collection.

57 Williams, "Where Georgia Was Born," undated, 955, envelope j, Item XXVI, Box 5, Williams Collection.

people there remind him too much of African-Americans to be aesthetically pleasing. "Reputedly lighter in tone than most Malays (or shall I say Netherlands Indians?) the people look more like negroes because their way of wearing clothes has a low-class Harlem touch. And when I say 'Harlem' I don't mean 'Haarlem.'"[58]

Williams's images of African-Americans appear to have been influenced by minstrel shows, with their characterizations of black men as foolish ineffectual dandies.[59] Williams was familiar with this form of entertainment, for not only did he watch minstrel shows, which were performed overseas in American clubs, he also may have performed in at least one himself. Staying in Constantinople, he recorded on 2 April, 1929: "Minstrel Show and Dance for the Officers of the 'Raleigh.' Reception on board. Dinner at C.C. [probably Constantinople Club] with Miss Ryan. Home late. Skin on face all worn off." The previous day, there had been another minstrel show, duly recorded, at the Union Francoise.[60]

Skin tone was an important marker in Williams's taxonomy of human groups, but so was the even more aesthetic-dominant category of clothing style. In a letter to his wife, Williams jocularly complained that, "When the ship touched port in New Guinea the natives came rushing out to modernize themselves. A few more cargoes of polo shirts, football shoes and felt hats and where will the poor photographer find subjects he couldn't find in Harlem."[61] Despite their lack of relationship to Americans of African descent, Williams implies, with Western clothes the dark-skinned, frizzy-haired New Guinea islanders are visually interchangeable with African-Americans, a circumstance unfortunate not only for the photographer, but – at least as this photographer is concerned – for the New Guineans as well.

Interestingly, Lutz and Collins find *National Geographic* making similar sorts of negative equivalences 50 years later; in a 1989 editorial board meeting, editors considered an article of the Efe in central Africa to be exchangeable with an article on East Harlem in New York City – and judged it undesirable to run both in the same issue.[62]

Williams's inability to recognize that his condescension of black people, particularly African-Americans, was inconsistent with his often-expressed humanism in no way disqualified him as an appropriate seeing-man for the *National Geographic Magazine*. As noted in Chapter 2, National Geographic treated African-Americans with disdain both in the magazine and at the Society. In 1910, for example, the *National Geographic Magazine* published an article that noted approvingly that "the negroes of Liberia are as polite and respectful to the white man as they are in Kentucky."[63] In a 1937 article on Mississippi, the caption for a photograph of rural

---

58 Ibid.

59 See, for example, Marlon Riggs' documentary film *Ethnic Notions* (San Francisco: California Newsreel, 1986).

60 Williams, Line-a-Day Diary, 2 April 1929, Williams Collection.

61 MOW to DWW, 15 April 1937, Williams Collection.

62 Lutz and Collins, *Reading National Geographic*, 50-51.

63 Edgar Allen Forbes, "Notes on the Only American Colony in the World," *NGM* 21 (September 1910), 723.

African-Americans sitting in a mule wagon informs readers that "'Sadaday' night is traditional 'darkey night' up-State. Then whites stay off the streets and the black families in pre-Sunday best emerge from 'catfish rows' and hold orderly carnival."[64] Four of the five adults are looking at the camera, which is slightly behind them to the right, and they look somewhat apprehensive. Juxtaposed with their expressions, the caption's jocular, teasing tone suggests Pratt's critique of John Barrow's eighteenth-century coding of !Kung communities' fear-based behavior of hiding by day and emerging at night as cultural "custom."[65]

Cramped style and nervous editors

For the most part, Williams delivered the goods that the *National Geographic Magazine* wanted and needed. And for the most part, he was happy with his arrangement. But Williams was not always pleased with the way the Geographic was run and marketed. His disagreements with his editors over the *National Geographic Magazine*'s style and intentions both trouble and reaffirm the position of "seeing-man" that Williams otherwise filled so well. At times, Williams experienced bouts of self-consciousness over the embedded nature of his work. That these doubts went unpublished, and that he nearly always – with one major exception – acceded to his editors' wishes, allowed for the perpetuation of the embedded "seeing-man" narrative.

Maynard Owen Williams's periodic tussles with *Geographic* editors – generally Grosvenor and the magazine's second-in-command, John Oliver La Gorce – are evident at least from the early 1920s, and centered mostly on what could be deemed stylistic issues. Williams's prose leaned towards the corny and sentimental, peppered with literary, Biblical and historical allusions. Editing tended to prune away his allusional tendencies, a process that caused both Williams and his editors dismay.

Editorially, Williams's biggest problem lay in what may be described as a kind of passive-aggressive elitism. Complaining about "the many inhibitions which writing for the Geographic imposes," Williams protested to La Gorce that "inhibitions of all kinds cramp my style, not my 'high-brow' manner but my natural style."[66] Editors, however, did their best to extract the "highbrow" from Williams's "natural style." Relatively early in his *National Geographic* career, Williams had suggested doing an article on "trailing Henry James through France," captioning current photographs of France with James quotations from 50 years earlier.[67] The idea fell flat at

---

64 J. R. Hildebrand, "Machines Come to Mississippi," with photographs by Joseph Baylor Roberts, *NGM* 72 (September 1937), 271. "GOIN' TO TOWN" is the caption header, and the caption continues: "This family is headed for Holly Springs, starting early for the front-yard rummage sales held there weekends, in which castoff clothing and other goods are spread on garden fences." Hildebrand was assistant editor of the magazine at the time.

65 Pratt, *Imperial Eyes*, 64.

66 MOW to JOL, 24 August 1923, NGS Records Society.

67 Ibid.

headquarters. Throughout his tenure at the *Geographic*, Williams was periodically reprimanded for being too erudite and assuming too much education on the part of the readers. "Please remember always to make your text as simple and natural as you can – so simple that a child of ten can understand it. Avoid indefinite allusions that the ordinary reader cannot grasp," Grosvenor wrote to Williams in 1930.[68] "Why do you spoil a good story by attributing to readers knowledge that only 1 in a million has," Grosvenor reiterated eight years later. "Please remember that simple language, direct, clear, easy to understand words and references are what Nat Geog Mag [sic] readers want. These persistent insertions in your articles make your editor nervous and uneasy and always afraid to pass your copy for printing until he has personally reviewed it."[69]

Editors also pressed Williams to provide details in his stories along the line of what Howard Abramson called that "maddening stream of disconnected facts, so typical of National Geographic features."[70] From one editor, Williams received "a very friendly letter asking me to write a story so cram full of detail and facts that it sounds like a Sears-Roebuck catalogue and that's what my next story will be." Again presenting himself as more artist than scientist, more subjective than objective, Williams admitted, "I lose some of the Sears-Roebuck specifications while striving to let a photograph speak the message."[71]

Another prickly thorn for Williams in his relationship with the *Geographic* was color photography. "I was afraid that the color difficulties had you licked," wrote Grosvenor in 1935. "Good luck to your continued color experiments and please remember whether we are 20 or 50 or 70, we must keep the open mind and be ready 1) to learn, 2) to adapt our methods to changing conditions and 3) to strive constantly to improve our technique… ."[72] On a purely technical level, Williams expressed a clear preference for working with black and white. For one, the color processes in the 1920s and 1930s – there were several competing versions – required more light and more equipment than did black-and white.[73] As Williams noted in a 1927 article, "color photographs require sixty times the amount of light of ordinary portraits."[74] Color photographs were generally landscapes and portraits: stationary subjects. Williams liked using the Graflex camera, which was easily portable and unobtrusive and good for populated photographs of markets and festivals. "[M]uch against conditions that hamper such work, I have worked hard on color not because I like it but because Dr. Grosvenor and the Staff do – and want it," he wrote to Daisy.[75]

---

68 GHG to MOW, 20 February 1930, Item XLII, Box 10, Williams Collection.

69 GHG to MOW, 10 May 1938, Item XLII, Box 10, Williams Collection.

70 Abramson, *Behind America's Lens*, 161–2.

71 MOW to JOL, 24 August 1923, NGS Records Society.

72 GHG to MOW, 21 May 1935, Item XLII, Box 10, Williams Collection.

73 Vesilind, *Development of Color Photography*.

74 Williams, "Color Records From the Changing Life of the Holy City," *NGM* 52 (Dec 1927), 682.

75 MOW to DWW, 21 November 1937, Williams Collection.

Williams's elitist and artistic inclinations were another source of his problems with color photography. He found himself annoyed at the kind of color photographs his editors wanted. "Mister Turner would have a fine chance of selling a sunset or a Venetian scene to the Geographic," he reflected in 1937. "'Might be anywhere,' re the matchless dome of St. Marks. 'Never saw a sunset like that in your life' – and too, too true. Most of us haven't. We see what the chromo publishers picture, not what Turner revealed to those who having eyes, see, once he's shown them the way." Alluding to the *Geographic*'s demand for a certain kind of color photography, Williams continued, "It is not well to count one's plates before they are matched to the idea that color must – to be worthy of reproduction – be color with a capital Sea, a couple of Ohs, a wholly silent 'Arh!' and a 'L' of a lot of it."[76]

The contrast Williams makes between the soft tones and subtle glow of Turner's paintings and the saturated color of *National Geographic Magazine* photographs, together with his repeated attempts at allusional prose, suggests a concern beyond that of style or technique. While, by the 1920s, Grosvenor's generic reader was essentially middle-class and middle-brow, Williams continually assumed a readership culled from his own genteel and educated milieu. Williams saw his work, and ideally, the *Geographic*'s, as part of a certain cultural realm the prevailing referent of which was a highly prescribed and elite liberal arts education.[77] Both Williams and Grosvenor were raised and educated in this cultural realm.[78] Grosvenor was interested, however, in steering his magazine into a truly popular product, *popular* in both the sense of being well-liked by many people and appealing to a broad range of people, "the masses." Grosvenor spoke of *National Geographic*'s envisioned reader as "the lonely forest ranger, the clerk at his desk, the plumber, the teacher, the eight-year-old boy or the octogenarian."[79]

Although Williams was proud of his ability to develop friendly and respectful relations with people from all economic or social backgrounds, and was certainly less elitist in his daily life than Grosvenor, he nevertheless wished that the *Geographic* had a more educated readership. His letters to people both inside and outside the Geographic reveal the extent to which Williams – who also had a journalism degree from Columbia – struggled against his editors' efforts to "dumb down" the magazine in pursuit of a wider circulation. "I have long wished that the National Geographic Society would limit its membership to 2,000,000," Williams wrote towards the end

76 MOW to DWW, 17 October 1937, Williams Collection.

77 In the early twentieth century, such an education would have comprised religion (usually specifically Protestant Christianity), literature, music and the arts, history, classical studies and languages, modern foreign languages – generally German or French, and a small but increasing smattering of sciences and mathematics. Recipients of such an education primarily shared European descent and presumably would be able to devote leisure time to cultural and educational activities.

78 The senior Grosvenor taught at government and politics at Amherst College in Massachusetts; the senior Williams taught Latin and Greek at Kalamazoo College in Michigan.

79 Quoted in Pauly, "World and All," 530.

of his tenure at the Geographic. "In a society which lets the 'Atlantic' and 'Harper's' almost die out, 1,800,000 Americans plus 200,000 relatively needy foreigners is enough. Then the editors could publish for a select audience rather than dangle cheap bait in front of a group which would inevitably lower the general standards."[80]

Insidious 'peeping Toms to nations'

Williams's elitism is evidenced as well in his attitude toward tourists and tourism. Williams resented tourists for several reasons, not the least being that their swelling numbers threatened to include him – an American with a camera – by association within their superficial ranks. He also saw tourism as an unwanted medium of social change, bringing unpleasant aspects of capitalism to places that were, Williams believed, better off without it. To a considerable degree, Williams's responses to this change were framed in aesthetic, rather than political or economic terms. Still, he questioned the familiar line of social evolutionary progress that the Geographic peddled, and interrogated his own role in a project that seemed to undermine that which he cherished.

Throughout his adult life, Williams reflected on travel as vocation and avocation. As a young man writing for the *Christian Herald*, Williams had expressed his disgust for tourists in a revealing taxonomic triumvirate of travelers. In the first category, which Williams declared he honored, was the scholar, who

> searches the records of the past, translates mouldy manuscripts or reconstructs the life of a race from a piece of pottery or a bronze safety pin… .
>
> Another man circles the earth and leaves friends and country to search for living knowledge. He studies the unity in variety and proves to the veriest snob the brotherhood of man… . I admire him, for he has a vision beyond the circle of self and the boundary of one nation.

While Williams respected the first category of the scholar-scientist traveler, it was the second category, that of the outward-thinking spiritual traveler, concerned with contemporary geographical connections between individuals rather than historical relationships, to which he aspired. It was within this framework that he signed on in 1919 as one of the *National Geographic Magazine*'s first staff writer-photographers.

While Williams placed himself and the *National Geographic Magazine* in the altruistic setting of the second category, he found that mode threatened by the increasing number of tourists. The tourist, that third category of traveler,

> travels for amusement and thrills. The indecency of such a person, lifting a cover off a section of the world to gratify personal curiosity, is as great as is that of an idle spectator at a surgical operation. If he thought poison would make the "native" more interesting,

---

80 MOW to Weimar K. Hicks, 7 October 1954, Trustees Folder, MOW 1953–55, Weimar K. Hicks Papers, Kalamazoo College.

> he would give it, slowly. We have no more right to be "peeping Toms" to nations than we have to individuals… . If they dress differently, act differently or think differently the tourist is satisfied, for that gives him something to criticize and act superior about.[81]

Williams's disgust is remarkable considering how close to his future vocational home his description struck. He marked a fine line between the admirable someone who "circles the earth … to search for living knowledge" and the despicable someone who travels "to gratify personal curiosity," and Williams's early description of the offensive desires of the tourist does not sound terribly different from the *National Geographic* project. Throughout his tenure at the *Geographic*, Williams had to continually face that he was part of an enterprise whose very success was based on the infusion of what Williams earmarked as tourist sensibilities: "peeping Toms" to nations, desiring difference, preferring the colorful superficiality of sheer display to the depth of historical context. After all, as a *National Geographic Magazine* photographer, Williams did what he could to "make the 'native' more interesting," such as having them wear – or not wear – particular forms of clothing that were not their usual attire. And as with the tourists he so virulently criticized, he also ended up complaining that some places – and some people – were "unpicturesque."[82]

Over the years, Williams became increasingly concerned about the effects of tourist contact on the places and people they traveled to see. Tourism brought contact with that Western-cosmopolitan complex, as Williams saw it, of consumerism/capitalism/modernity, introducing contemporary Western fashions, consumer concerns and economic exploitation to the previously tradition-bound and materialistically "innocent" lands and people. "Glory be, I landed in Cyprus before the first cruise ship stopped there," Williams wrote to illustrations editor Franklin Fisher in 1927. "I shall regret the spoiling that must come."[83] His concern was part personal – he wanted to beat the crowd of tourists so that he could experience an "unspoiled" place – and part professional – he could still photographically capture the "unspoiled" essence of the place.

In a similar rumination ten years later, Williams put himself and the *National Geographic Magazine* in the center of the process he found so reprehensible. Writing to his wife from Singapore in 1937, he discussed a conversation he had with a ship's captain about "unspoiled lesser" islands.

> "Then the tourists will come and the place will be spoiled. Soon there will be no unspoiled places on earth," says [the captain]. And I, traveling in this odious business of directing vulgar tourists to places that would be better off without them, get a reduction of 25% because I'm a writer, the K.P.M. and The Geographic conspire against the captain and me making capital of the exploitation of people who can't help themselves … [84]

---

81 Williams, "Tourist Stuff" manuscript, p. 50, envelope a, Item XXVI, Box 5, Williams Collection.

82 MOW to DWW, 25 March 1937 and 9 September 1937, Williams Collection.

83 MOW to Franklin Fisher, 14 May 1927, National Geographic Records Library.

84 MOW to DWW, 31 March 1937, Williams Collection.

Usually successful at suppressing and ignoring the feeling that he worked for an organization that operated against what he believed, Williams reveals here his ongoing struggle over the contradictions of that arrangement.

Although he was still prone to aestheticizing social conditions, by the 1930s Williams perceived capitalism as a global system being foisted upon places and people ever-more remote from the cosmopolitan Western center. Never fond of the "idle rich" to start with,[85] the Depression pushed him towards a more cynical view of capitalism.[86] Williams had lost some money in the crash, and he was forced to take a pay cut at the Geographic as well; membership suffered in the 1930s when even armchair voyages became a luxury.

Like many of the contributors to the *National Geographic Magazine*, however, Williams was more concerned about the cultural imperialism of the West than he was about European and American systematic exploitation of the labor and resources of the colonized world. For Williams, capitalism was a *cultural* product of the West, a package deal that included movies, brassieres, Fedoras, and corrugated iron roofs together with incorporation, at a disadvantaged position, into a competitive global economic system. Within *National Geographic*'s narrative of progress, capitalism was nearly synonymous with modernization, and modernization was inevitable.[87]

Williams also swept Christianity into the melange of generally Western influences threatening to contaminate the world's pure and creatively differentiated cultures. He grew increasingly unhappy with what he saw as the ideological interplay between "American" and "Christian," with "American" often implying inculcation in global capitalism. "When these educated young Christians get to the industrial age and wage slavery, I wonder how much comfort Christianity will be," he wrote to his wife regarding "the Christianized Bataks."[88]

But concerns about wage slavery generally took a back seat to Williams's chronic consternation over the more visible manifestations of cultural change, the aesthetics of clothing and architecture. His letters express irritation with missionaries more for encouraging the inhabitants to wear ugly Western-based clothes than for interfering with their traditional spiritual beliefs. Similarly, he blamed colonialism for making

---

85 At the time of Williams's "Tourist Stuff" piece, wealth was nearly a prerequisite for travel, which only seemed to make the category "tourist" that much more dislikeable. Williams particularly attacks "that being whose money and ennui run about 182½% and whose brains and real work in life make up the 200." "Tourist Stuff," 49, envelope a, Item XXVI, Box 5, Williams Collection.

86 Indeed, as discussed in Chapter 2, Williams was an initial supporter of the "Soviet experiment." He spent considerable time traveling in Russia and the Soviet Union in the early revolutionary days, and tried to convince a strongly anti-Soviet Grosvenor to run articles about the country.

87 See also Lutz and Collins, *Reading National Geographic.*

88 MOW to DWW, 4 June 1937, Williams Collection.

the built environment ugly with such elements as corrugated iron roofs and military bulwarks. He seemed less perturbed about colonial exploitation.[89]

To some extent, Williams's aesthetic "appall-o-meter"[90] was guided by his elitism. Williams disliked what he saw as Western aesthetic encroachments both because it erased previous cultural uniqueness, and because what was imported was often low-class. Western-style clothes distributed by missionaries were usually the clothes that would adorn either the poor of North America and Europe or had been thrown away. While corrugated iron roofs were Western products, so too were slate roofs, but it was the rough metal that covered many modest abodes.

Both within the magazine and in the outside world which he was employed to capture, Maynard Owen Williams shunned the vulgar and cheap. Tackiness was not restricted to cultural trappings or unappealing, low-class sights, but encompassed moral vulgarity as well. For Williams, the Geographic's high-minded template of portraying a friendly world provided a bulwark against baseness.

'This world is not a monkey cage'

Thirteen years after his article on the 1931–32 Citroën-Haardt expedition, in which Williams spoke of himself as a "conscientious photographer" who cultivated mutually friendly relations between photographer and subject, he found himself again defending his position in the name of "world understanding." In 1945, Williams received a copy before its official release of a memo for field photographers written by staff at the *National Geographic* home office before its official release. It read: PHOTOGRAPHERS: WRITE FULL CAPTIONS FOR YOUR PICTURES! The memo went on to tell photographers to be as explicit and detailed as possible in providing the *what*, *where*, *when*, and *why* of each photograph in the full sets of Kodachromes and contact prints required from an assignment.[91]

It is striking that the first question of journalism, *who*, was left off the litany. Human subjects presumably came under the rubric of the *what* question: "Tell briefly what is included in the photograph. If it is something strange, curious or unusual, make notes on the spot from an authority (keep his name and address) who knows the subject." Further down in the memo, photographers are told that "Names of plants, birds, animals or other subjects should be stated – generalities are not wanted." Names of people are not among the specifics requested by the picture editors. Indeed, humans as photographic subjects are first mentioned in the memo about halfway through, with the suggestion that: "How primitive models behave

89 In his sojourn through the Netherlands Indies in 1937, for example, Williams remarked disapprovingly on Dutch disrespect for indigenous islanders, commenting slightly more frequently and approvingly on Dutch efficiency. He seemed most outraged by the effect colonialism had on air mail routes and costs: It was considerably more expensive and difficult to send mail to the U.S. than to the Netherlands.

90 Appellation credit to *In These Times*.

91 Item XLII, Box 10, Williams Collection.

under the camera – whether bashful, angry, or sullen – often makes interesting reading."

Williams did not receive the memo calmly – it gave him "some bad hours."[92] After thinking "it over for quite a while"[93] he wrote an impassioned letter to his "Dear Old Friend" Melville Grosvenor, Gilbert Grosvenor's son, who was then a *Geographic* picture editor. Over the next few days, he fired off similar letters to others on the editorial staff.[94] Generally, Williams favored praise and acquiescence over criticism in his *National Geographic* correspondences, particularly with his top "bosses," editor-in-chief Gilbert H. Grosvenor and John Oliver La Gorce. But his letters in response to the memo burn with disbelief and anger. He wrote to Melville Grosvenor, "that scarlet sentence is a betrayal of a great ideal – that of contributing to world understanding, not of a wild man of Borneo by moronic gawpers but as Man to Man." He continued:

> This world is not a monkey cage, full of "primitives" who must perform, however "angrily or sullenly" for a peep show mentality, nourished by tabloid sensationalism and held by uncomprehending dullards who consider themselves The Master Race.
>
> I don't believe you can show me one angry or sullen face among the thousands of negatives it has been my happy privilege to make.
>
> I have no more right to subject a "primitive model" to such treatment than he has to torture my wife –or yours. I don't think that such instructions or suggestions are fair; to the photographers who risk their lives to the Society, which risks its reputation in such questionable procedures.
>
> Have we led a million members toward a better understanding without attaining it in our own instructions to those who represent The Society among the peoples of the world?[95]

The memo calls into question the *Geographic*'s "happy humanist" practice. But for Williams, it was unimaginable that *National Geographic*'s visual friendliness was simply a matter of style or mere veneer; it was a mission to which he had served for more than 25 years. "I don't know one of us who does not take pride in the thought that he is serving a noble and high-minded organization," he told Gilbert Grosvenor. "But we don't make 'primitive models' submit to such indignities that they become sullen or angry. I think our pictures show that fact."[96]

The relationship between the photographer and the photographed had long been important to Williams, and in his *Geographic* articles he regularly commented on his interactions with people he photographed. In his discussion of photographing the Chinese cobbler, Williams spelled out his interpretation of the reciprocated "Man to Man" relationship: "He and I cooperated in that picture through a tacit understanding.

---

92 MOW to GHG, 7 April 1945, Item XLII, Box 10, Williams Collection.

93 Ibid.

94 Recipients included Gilbert H. Grosvenor, assistant editor J. R. "Hilde" Hildebrand, and Edwin "Bud" Wisherd, chief of the photo lab.

95 MOW to Melville Bell Grosvenor, 5 April 1945, Item XLII, Box 10, Williams Collection.

96 MOW to GHG, 7 April 1945, Item XLII, Box 10, Williams Collection.

In this limited sense, we like each other."[97] Presumably, this "tacit understanding" involved the cobbler's consent to be photographed, but Williams pushed his interpretation further: "we like each other." Stressing cooperation, understanding and mutual feeling between himself and his subject, Williams made an effort to keep the relationship on an even, reciprocated level, the subject as much an agent in the process as the photographer. The act of taking the photograph, and its publication in the *National Geographic Magazine*, stood for "evidence of mutual friendliness,"[98] evidence that, for Williams, only served his purpose of "world understanding."

While Williams conceivably might have made the same argument about photographing a woman, it is telling that his argument for mutual friendliness was based on a male subject. One of the striking things about Williams's response to the 1945 memo is the markedly gendered framework in which he places the *Geographic*'s work. Whatever the memo's gendered connotations of "primitive" might have been, for Williams, the so-called "primitive" referred to is an adult male. The relationship Williams understands as the heart of the *Geographic* experience, that between the magazine's readers and the photographers' subjects, is not that of the demeaning and unilateral viewing of "a wild man of Borneo by moronic gawpers" but the parallel and literally reflective "Man to Man." Williams's belief in *National Geographic* hangs on that reciprocated vision of the man in the mirror. Rather than a direct two-way relationship, however, the mirror is mediated, enabled, catalyzed by a third masculine being, the "field man." Williams is this pivotal third man. Although it is clear that Williams's objection to the sentence is based on its dehumanizing of the intended subject, his identification of the (genderless) "primitive" and the (genderless) "reader" with men betrays not only a normative masculinization of the generic, but Williams's own identification with the photographed subject. Perhaps it is this identification that causes such cognitive dissonance for Williams. The objectification does not just demean the subject, but is specifically emasculating.

Williams drives home this violation of the "primitive" subject by the photographer through an analogy between that situation and physical violence against a respectable white American woman: "I have no more right to subject a 'primitive model' to such treatment than he has to torture my wife – or yours." In his letter to Bud Wisherd, head of the *Geographic*'s photo lab, Williams similarly asserts: "If such malpractice 'makes interesting reading' so does rape."[99] Clearly, the "primitive" man is in the woman's position, raped by the violating photographer. The equal relationship of "Man to Man" has been replaced by "Man to Woman," the Man being the *Geographic's* readership and the intervening photographer, and the Woman being the photographed subject. "Man to Man" has a meaning about power relations that "Man to Woman" or "Woman to Woman" does not.

---

97 Williams, "Mediterranean to the Yellow Sea," 560–61.

98 Ibid.

99 MOW to Edwin L. "Bud" Wisherd, no date [April 1945], Item XLII, Box 10, Williams Collection.

Women, of course, were frequent subjects of *National Geographic Magazine* photographs. Yet despite 25 years of chasing women's smiles, Williams frames the subject as exclusively male. Williams is not simply using the male as generic, as was the prevailing conceptual and linguistic use at the time. Rather, it is that objectification of women, particularly those deemed "primitive," is normative. There is little jarring for Williams about objectifying women, while objectifying a man feels wrong and uncomfortable, despite his "primitive" culture or nature.

Another reason for the erasure of women from Williams's photographer-subject relationship has to do with Williams's belief that through its magazine, the National Geographic Society engaged in the basic political work of furthering world brotherhood and friendship between nations. He saw the *Geographic* as a diplomatic mediator. Within this political framework, the actors – the readers and the photographed subjects – are adult males, the franchised in the arena of world citizenship. World *brother*hood is to be achieved through a social contract between men.[100] Because the "primitives" of the memo are rendered female and objectified, they are unfit to participate in the negotiations of world brotherhood.

Williams saw a version of world brotherhood in the National Geographic Society, making note in his field notes and letters each time he met a resident of the country he was visiting who was a member of the Society.[101] The Geographic's foreign membership, as judged by Williams's notes, consisted primarily of elites and educators. Williams held this rendering before Gilbert Grosvenor when condemning the memo. "All around the Persian Gulf are gentlemen whom a careless person might regard as 'primitive models' and they are eager to belong to the Society," he told Grosvenor. "They would not if the Society made them angry or sullen."[102]

Williams was probably not the only Geographic staffer to object to what he called "the sadistic element."[103] The offending sentence in the memo was removed before it was officially distributed. A photographer's guide enclosed with a letter dated 1952 makes no mention of "primitives," and aside from telling prospective contributors that "releases are not required from people appearing in photographs for editorial use," the only suggestion regarding the expressions of possible subjects is that "smiling faces, when appropriate, also tend to brighten a picture."[104]

---

100 As Carole Pateman notes, the social contract – first developed, as explained by Rousseau, between "primitive" men – is explicitly structured so as to exclude women; women are negotiated about, not with. See Carole Pateman, *The Sexual Contract* (Stanford: Stanford University Press, 1988).

101 For example, Williams noted in the few lines allotted in his diary that the "son of the headman" in Boghaz Kerir "admires the Geographic" (13 October 1928, Line-a-Day Diary, Williams Collection).

102 MOW to GHG, 7 April 1945, Item XLII , Box 10, Williams Collection.

103 MOW to J.R. Hildebrand, 9 April 1945, Item XLII, Box 10, Williams Collection.

104 *National Geographic* Records Library.

Figure 4.6 "The Face with the Smile Wins"
(Maynard Owen Williams/*National Geographic* Image Collection)

'My feeling is for the beautiful'

Mary Louise Pratt's figure of "seeing-man" is based on a particular variety of Victorian traveler, but the concept remains a useful one in examining *National Geographic* and its contributors. For one, under Gilbert H. Grosvenor's tutelage, the Geographic held onto styles of travel writing – and the practitioners of such writing – that had been developed in the nineteenth century, causing commentators through the twentieth century to refer to *National Geographic* as "old-fashioned." But the *National Geographic Magazine* also brought "seeing-man" into the twentieth century, expanding the scope of his vision through the heavy use of photography.

In many respects, Williams made a perfect "seeing-man" for the Geographic. Whether with delight at "natural" exposed brown breasts and "true" ancestral costumes, or disdain for the "incongruously" clad Asian/Pacific inhabitants in Western garb, Williams framed the world in aesthetic terms. As a photographer, he aspired to emulate his favorite landscape painters. "There is one challenge eternally before me when I'm in the East – to put the scene into the proper words or to picture it with its proper atmosphere and sense of values," Williams mused to his wife in 1937. "Grant Wood feels that challenge in Iowa. I feel it in Asia. He seeks to picture the real. My feeling is for the beautiful … ."[105]

Through his vision, he aimed to deliver the world to his audience, although it was, perhaps, more American and middle-brow than he would have preferred. In line with the institution to which he devoted himself for nearly 40 years, he sought to excise politics from his portrayals, going out of his way to capture images that would tell the story he wanted to tell of a beautiful and friendly world.

Still there was a difference between Williams's vision and that of the institution for which he worked. The smiling faces, so desired by *National Geographic Magazine* editors because they "tend to brighten a picture," represented a deep-seated belief for Williams. Although the end effect was basically the same – pretty pictures, smiling people, positive feelings – a close look at the interactions between Williams and his editors shows the active negotiations behind the stories and images the *National Geographic Magazine* presented to its readers. Williams's letters, unpublished manuscripts and even his published articles reveal internal negotiations as well. From early on, he worried that his work was fodder for tourism, which he saw as spreading some of the crasser elements of capitalist American culture. While he tended to ignore political and social dominance of colonizing powers, he saw and was troubled by cultural changes wrought by the colonizers, and by the impact of global capitalism in general. Williams's awareness of and conflict over his participation in the formation of an American version of "European planetary consciousness" allows us to more deeply examine the contradictions of the anti-conquest position, particularly within the context of American imperialism.

105 MOW to DWW, 30 August 1937, Williams Collection.

Chapter 5

Harriet Chalmers Adams: Intricacies of Class, Gender and Gusto

No woman wrote more articles for *National Geographic* in the magazine's first 50 years than Harriet Chalmers Adams. During her 30-year affiliation with the National Geographic Society, Adams contributed 21 articles to the magazine, mostly about Latin America and other places related to the Spanish empire. The magazine's top editors considered her one of their experts on South America, and her vision of the Americas cohered well with the *Geographic*'s.

Adams began her association with the Geographic through the Society's lecture series. Returning in 1906 from two years' travel in South America, Adams wrote a letter to Gilbert H. Grosvenor summarizing her experiences and offering to lecture at the Society. "The descriptive circular which she enclosed with her letter was so interestingly prepared and so unusual then, being illustrated with many beautiful photographs taken by Mr. Adams, that our organization invited her to come to Washington," Grosvenor recalled. "This was Mrs. Adams's first lecture, the National Geographic Society was her first audience and the *National Geographic Magazine* published her first article. All our members were charmed by her pleasing, human narrative and encouraged her to continue lecturing and writing."[1]

Her first *Geographic* article and lecture, in 1907, initiated a series drawn from her 1904–1906 South American travels.[2] Her missives sent to newspapers from the French front during World War I frequently identified her as *National Geographic*'s war correspondent and several of her stories were distributed by the National Geographic Society as the Society's Bureau of Geographic Information press releases. By 1925, John Oliver La Gorce was touting Adams as the person "whom we regard as the foremost traveler and writer in the United States. … She possesses a beautiful writing style and her facts we have found to be accurate and unusual; in fact we place every

1 GHG to the Society of Woman Geographers, 9 December 1937, Part I: Container 1, Records of the Society of Woman Geographers, Manuscript Division, Library of Congress, Washington, D.C. [Hereafter referred to as SWG Records].

2 Harriet Chalmers Adams, "Picturesque Paramaribo," *NGM* 43 (June 1907): 365–373; Adams, "The East Indians in the New World," *NGM* 43 (July 1907): 485–491. Also, Adams, "The Story of Adventures in Peru," 13 December 1907, Harriet Chalmers Adams scrapbook, Stockton Public Library, Stockton, California. Unless otherwise noted, all newspaper, lecture program, and correspondence references in this chapter come from one of the six full scrapbooks donated to the Stockton Public Library [hereafter referred to as SPL] by Franklin Adams.

confidence in her work."[3] Indeed, La Gorce treated Adams as *National Geographic*'s South America expert, asking her to review articles submitted to the magazine.

La Gorce was not alone in his accolades. Newspapers in the early 1920s referred to Adams as "one of the seven greatest world travelers"[4] and "America's leading woman explorer."[5] In 1913, she was elected as one of the first women fellows of Britain's Royal Geographic Society, and many other geographical societies throughout the United States and Latin America honored her with memberships.[6] As some organizations, like New York's Explorers Club, still excluded women from membership, Adams helped to start the Society of Woman Geographers, of which she became the first president in 1925.

In this chapter, I use Adams'ss career to examine issues of gender in the arenas of exploration and National Geographic, and focus on Adams'ss work to further examine the presence and presentation of global politics in *National Geographic Magazine.*

'First white woman at number 8'

Harriet Chalmers Adams liked to explain her adventurousness through her parentage, linking her explorations abroad with her family's life in the "Wild West" of California. Her father's family had emigrated from Scotland to Canada when he was a child, and in 1864, young Alexander Chalmers went to California to make his livelihood in mining. With his older brother, who had been a California gold miner since 1842, he eventually settled in Stockton, opening up a dry goods store and later becoming superintendent of a mine. Harriet's mother, Frances Wilkins, also came from a California gold-mining-turned-merchant family, by way of colonial New Hampshire. Born in Stockton in 1872, Harriet Chalmers grew up swimming, hiking and horseback riding all over the San Joaquin Valley. She was schooled by private tutors, but it was her father who taught her to hunt, fish and ride. "Before I was fourteen, I had completed, on horseback, the exploration of the entire Sierra Nevada Mountain chain with my father," she told one interviewer.[7]

3 Copy of letter to George Putnam of Putnam's Sons publishers enclosed with letter from JOL to Harriet Chalmers Adams [hereafter HCA in reference to letters and journal entries], 11 June 1925, SPL.

4 Charleston *Daily Mail* [reprint from New York source], 12 December 1920, SPL.

5 *The Herald*, 12 March 1922, SPL.

6 While most other geographical societies in the British Empire and the United States had accepted women as either members or honorary members since their founding, the RGS, after abortive efforts in 1892–3, did not open its membership to women until 1913. Adams's election to RGS fellowship came a month after the election of American Fanny Bullock Workman, known for her Himalayan surveys. See Morag Bell and Cheryl McEwan, "The Admission of Women Fellows to the Royal Geographical Society, 1892–1914; the Controversy and the Outcome," *The Geographical Journal* 162 (November 1996): 295–312.

7 Elna Harwood Wharton, "A Woman Turns Geographer," *The Forecast* (July 1930), 25, SPL.

In 1899, Harriet married Franklin Pierce Adams, a fellow Californian. Four years older than Harriet, Franklin Adams came from another of Stockton's "fine" families, and following stints as a theater manager and a newspaper publisher, was working as an electrician at the Stockton Gas and Electric Company at the time of their marriage. The couple soon embarked on their career of travel; following their marriage, they "motored" all over California and Mexico.

Adams's first extensive trip, to Central and South America in 1904–1906, launched her career as an explorer. Although she accompanied her husband to South America at least partly under the auspices of his employment with the Inca Mining and Rubber Company, much of the trip was at the couple's own expense.[8] They went just about everywhere, crossing the Andes by horseback and mule, and canoeing down the Amazon. At times they contended with extremely rough conditions, walking through one river with water up to their shoulders and crossing elsewhere by crawling on a wet log, and they confronted rainstorms, blizzards, cold, hunger, lack of sleep and vampire bats.[9] Adams was conscious of her multiple outsider situation, noting in one diary entry, "I was first woman black or white to walk from La Union to West River and first woman there unless one counts the Savages. Also first white woman at Number 8 and La Union."[10]

Still, at the mining camp Adams rejoiced in finding a hot bath, drinks, electric lights and a comfortable bed.[11] They also spent many weeks in port cities and provincial and national capitals, where Harriet would study books and maps of the next leg of their journey and where they refreshed themselves with social dinners and cocktails. Both Adamses took photographs, and they must have carried a considerable amount of photographic equipment and accessories, for they produced – and developed themselves – hundreds of photographs and dozens of reels of movie film. On their return to the United States, Harriet had plenty of material to launch her new career as writer and lecturer. With or without her husband, but usually with, Adams made several more trips to South America, and established herself in the public eye as an expert on the region.

Fairly fluent in Spanish and Portuguese, and speaking French and German as well, Adams attributed her interest in Latin America to her childhood in the former Spanish colony of California. In the course of her life, she deliberately and methodically traveled to every place that had ever been part of Spain or its empire,

---

8 Adams's journal entries from this period (January 1904–May 1906) suggest that Franklin Adams was a mine inspector. Enormous thanks to Kate Davis for giving me her transcription of these journals, which Adams's grand-niece, Nancy Ditz, graciously had lent her. Davis's masters thesis was the first substantial work on Harriet Chalmers Adams produced since Adams's death. M. Kathryn Davis, "The Forgotten Life of Harriet Chalmers Adams: Geographer, Explorer, Feminist" (Masters Thesis, San Francisco State University, 1995). Durlynn Anema's young adult biography, *Harriet Chalmers Adams: Adventurer and Explorer* (Aurora, Colorado: National Writers Press) came out in 1997.

9 Diary entries April and May 1904.

10 Diary entry, 4 May 1904. The area in which she was traveling is located in Peru near the Bolivian border.

11 Diary entries, 23 April 1904 and 24 April 1904.

including the Philippines and North Africa. She was fascinated by Spain's colonial history, and suggested to Gilbert Grosvenor a series of articles tracing the paths of several conquistadors;[12] in 1912 she retraced Christopher Columbus's travels from Spanish towns and cities to North America.[13] Adams'ss exploration of Haiti, part of the Adamses' travel on the island of Hispaniola, resulted in the capture of six solenodons, "rare antediluvian ancestor[s] of the rodent family," which they gave to the New York Zoological Society (Bronx Zoo) and the National Zoo in Washington, D.C.[14]

Both Harriet and Frank took photographs, although Franklin Adams got only a handful of photo credits early on. Some of the photographs that illustrate her articles are credited "By Harriet Chalmers Adams," while more are "From Harriet Chalmers Adams." Her early photos, confined by the camera technology of the period, tended toward the stiff and posed. Captions sometimes proclaimed the subjects "types." By the 1920s, though, Adams'ss photographs had developed more life and personality. Some subjects, such as a woman trolley conductor or a woman peddling handmade lace to a train passenger, are close within the frame but they seem oblivious to the camera's intrusion. Later captions, reflecting changing styles at the *Geographic*, were more detailed.

Adams took pride in her work for the *Geographic*, and she took care to give her editors the work they wanted. "[W]hat I would like most on earth to do is to accomplish work which would reflect glory on the National Geographic Society, which has so befriended me," she told Grosvenor.[15] The relationship between Adams and *National Geographic* proved mutually beneficial to them both. Her 1920 piece on Rio de Janeiro, Grosvenor said, was often "described as the ideal travel article."[16]

The Geographic used Adams in a new venture during the First World War. In 1916, Adams went to the French front as a war correspondent, the first woman correspondent permitted to do so, and one of the few people allowed by the French government to take pictures of actual battles.[17] Her French press pass said that she represented *Harper's Magazine*,[18] but as far as many American newspaper readers were concerned, she was the National Geographic Society's war correspondent.

As a monthly, however, the magazine could not be a purveyor of the latest news. Sensing a need – or a niche – Grosvenor initiated daily news bulletins for newspapers, and released them under the auspices of the Society's newly formed Bureau of Geographic Information. Adams became a leading bulletin writer, and the

---

12 HCA to GHG, 28 January 1926, SPL.

13 HCA scrapbook, SPL.

14 "Solenodon Paradoxus," *New York Zoological Society Bulletin* 41 (September 1910), SPL; William M. Mann, *Wild Animals In and Out of the Zoo* (Washington, D.C.: Smithsonian Institution, 1930), 230.

15 HCA to GHG, 29 March 1916, NGS Records Library.

16 Gilbert H. Grosvenor, HCA memorial manuscript, SWG Records.

17 Washington, D.C. *Star*, 26 September 1916, SPL.

18 Note typed by HCA in scrapbook, SPL.

Geographic's service distributed her bylined reports to 550 newspapers.[19] Some of the material from the bulletins reappeared in the single *National Geographic* article Adams published since 1910, "In French Lorraine: That Part of France Where the First American Soldiers Have Fallen."[20]

In addition to her articles for *National Geographic*, Adams wrote for a variety of other publications, including *The American Review of Reviews*, *The World's Work* (in Spanish for the Latin edition), *Tropical America*, and the *Ladies Home Journal*. She had book contracts with publishers such as Doubleday and Putnam, but never finished a manuscript, finding shorter articles and lectures more amenable to her busy schedule of travel and speaking engagements.[21]

Adams lectured widely, at geographical societies (American Geographical Society, Geographical Society of Philadelphia), private clubs (Detroit Athletic Club, University Club of San Francisco, Twentieth Century Club), colleges and universities (Vassar, Rutgers, Columbia), and in lecture series at such arenas as the Carnegie Lyceum, the Brooklyn Institute of Arts and Sciences (now the Brooklyn Museum), the Waldorf-Astoria, the Battle Creek Sanitarium, and Chautauqua. The J. B. Pond Lyceum Bureau represented Adams in the 1909–1910 and 1910–1911 lecture seasons.[22]

Adams was a Geographic loyalist, though, and proud of her connection to the Society. "Please note that I have put the National Geographic society first [in her promotional material] and that I value it more, and make mention of it oftener than of anything else connected with my work," she wrote to Grosvenor in 1914.[23] Both the La Gorces and the Grosvenors knew the Adamses socially. To the extent that they settled anywhere, Harriet and Frank made their home in a comfortable apartment in Washington, D.C. Between Harriet's affiliations with the National Geographic Society and the Society of Woman Geographers, and Frank's work at the Pan American Union, they had numerous connections to the capital's political-social-intellectual elite. From 1925 until her departure from the U.S. in 1933, Harriet Chalmers Adams presided over the Society of Woman Geographers, of which Elsie Bell Grosvenor, Gilbert's wife, was an early member. Frank was a member of the Cosmos Club, the private gentleman's club to which many of Washington's male educated elite – including Grosvenor and Maynard Owen Williams – belonged. La Gorce and Grosvenor were sufficiently friendly with the Adamses that La Gorce would occasionally address his notes to "Dear Hally" or "Hallie," though she seems

---

19 HCA scrapbooks, SPL; Abramson, *Behind America's Lens*, 182. No National Geographic history mentions Adams in any respect, let alone as a leading news bulletin writer.

20 Harriet Chalmers Adams, "In French Lorraine: That Part of France Where the First American Soldiers Have Fallen," *NGM* 32 (November–December 1917): 499–518.

21 HCA to Society of Woman Geographers Washington group, 3 August 1935, SWG Records.

22 HCA scrapbook, SPL.

23 HCA to GHG, 11 March 1914, NGS Records Library.

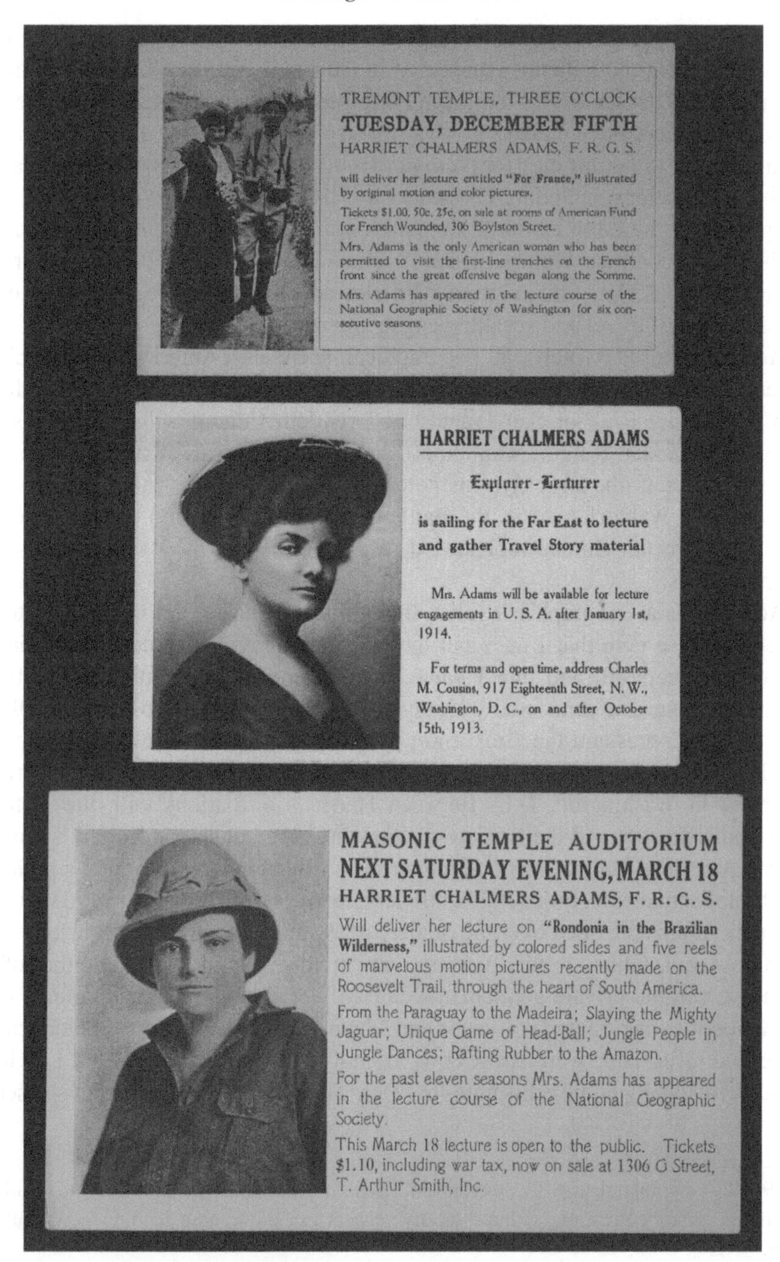

Figure 5.1 "You could not obtain a more popular, entertaining, or instructive lecturer than Mrs. Harriet Chalmers Adams"
(National Geographic editor Gilbert H. Grosvenor told Chicago's Chautauqua Institution)
(Society of Woman Geographers, Washington, D.C.)

to have been "Hellie" or "Helly" to those to whom she was closest.[24] Grosvenor considered the Adamses friends, and had them over for dinner.[25] And when the Adamses relocated to Europe in the mid-1930s, the Grosvenors sent them packages of current magazines such as *Time*, *The Atlantic*, *Asia*, the *Saturday Evening Post*, and the *Illustrated London News*, in addition to the latest edition of *National Geographic.*

Harriet Chalmers did not, however, get everything she wanted from the Geographic. Grosvenor refused Adams's request to have the initials F.R.G.S. (Fellow of the Royal Geographic Society) in her byline. The manager of the J.B. Pond Lyceum Bureau had asked her to display those illustrious initials, as had an editor for another publication, most likely *The World's Work*. "Both think it of value while few women are using it in these United States, and 'business is business,'" she explained.[26] The *Geographic* did sometimes publish authors' academic degrees or military affiliations in bylines, as with H.F. Lambart, B.Sc., D.L.S. and Lieutenant H. R. Thurber, U.S.N.[27] Grosvenor's refusal to print those particular initials was most likely based a real or perceived rivalrous relationship with "the royal of London." Adams tried a second time: "I use the F.R.G.S. after my name at times because it has helped me in placing magazine articles and I must help myself in this world of contest," she argued,[28] but Grosvenor did not relent.

Grosvenor also denied Adams funding for her exploration. Although the Adamses were of a "comfortable" background,[29] travel and photography were not cheap pastimes, and they relied on Frank's salary with the Pan American Union and Harriet's lecture and article fees.[30] In 1916, when Adams requested funding for an

---

24 Various letters, JOL to HCA, NGS Records Library; HCA memorial manuscript by Col. Charles Wellington Furlong, SWG Records. In the diary entry for 29 March 1904, she notes that "Helly" received 11 letters and "FPA" got 8.

25 Writing to his mother-in-law in 1909, Grosvenor noted: "Tonight we've just had a few friends to dinner, A. Henry Savage Landor [explorer who wrote about Tibet, Korea, and in a 1908 *Geographic* article, "Wildest Africa"], and Mrs. Harriet Chalmers Adams and her husband who have been everywhere in South America. She's the author of those beautiful articles in the Geographic Magazine." GHG to Mabel Hubbard Bell, 13 February 1909, Box 52, Bell Papers.

26 HCA to GHG, 11 March 1914, NGS Records Library. Adams's byline included F.R.G.S. on her *World's Work* article "Snapshots of Philippine America," May 1914: 31–42, as well as a decade later for "The Truth About Spain and Primo De Rivera," *The American Review of Reviews* 71 (January–June 1925): 69–72.

27 H. F. Lambart, "The Conquest of Mount Logan: North America's Second Highest Peak Yields to the Intrepid Attack of Canadian Climbers," *NGM* 49 (June 1926): 645–693; H. R. Thurber, "Collarin' Cape Cod: Experiences On Board a U.S. Navy Destroyer in a Wild Winter Storm," *NGM* 48 (October 1925): 427–472.

28 HCA to GHG, 29 March 1916, NGS Records Library.

29 GHG to the Society of Woman Geographers, 9 December 1937, SWG Records.

30 Their mid-1930s sojourn in Europe was largely precipitated by Frank Adams's feeling that they could not afford to live well in the U.S. on the meager income he received from his pension.

expedition she and her husband were planning to Saharan Africa in 1917,[31] Grosvenor replied with a letter acknowledging his interest in her proposed itinerary, particularly Kano, "which has long appealed to me as the center of the most interesting region in Africa, excepting, of course, Egypt." But he shifted the responsibility of her funding to the "Research Committee," expressing doubts that the committee would give her a grant. "Just at present we are conserving our resources with a view of some day building an auditorium next to our present buildings," he told one of his most favored speakers.[32] But, as Kate Davis reveals, during the period 1915–1919, the National Geographic Society funded seven expeditions to the Mt. Katmai region of Alaska. And in the year of Adams'ss request, 1916, the Society funded Frank M. Chapman's expedition to Peru's Urubamba Valley, where Adams had traveled 10 years before.[33] Instead of going on the planned African expedition, Adams went to France to cover the war.

It is difficult to know for sure how much the fact of Adams's gender played into the Society's refusal to fund her. Still, as Carolyn Bennett Patterson, who joined the magazine in the middle of the twentieth century, recalled, "The Geographic was, in fact, a very elite gentleman's club, which did not admit women."[34] During Adams's association with the National Geographic Society, from 1907 to 1937, only one woman, Eliza R. Scidmore, held any position among the editorial board, the Society's executive board, or on the Board of Trustees (originally the Board of Managers). Even after Adams's death in 1937, little of the Society's internal social structure changed during Gilbert H. Grosvenor's lengthy leadership. Patterson was hired in 1949 to be a library research assistant and eventually became *National Geographic*'s first full-time female editor, around 1965. "There was a very, very distinct idea of what ladies did and what ladies did not do here," she said.[35]

For much of the Geographic's existence, "woman" and "staff writer" or "editor" seemed to be incompatible terms of identity. Eliza Ruhamah Scidmore, who served on the Society's Board of Managers from the early 1890s into the first decade of the twentieth century, was an early and important exception. An established Washington, D.C.-based travel writer, Scidmore was part of an initial trio of associate editors (the other two were W.J. McGee and A.W. Greely) in the masthead's earliest days in 1896, and she remained in that position through 1903.[36] She contributed 16 articles, many illustrated with her own photographs, from 1894 to 1914.

---

31 HCA to GHG, 29 March 1916, NGS Records Library.

32 GHG to HCA, 30 March 1916, NGS Records Library.

33 Davis, "Forgotten Life."

34 Patterson, *Lands, Legends, and Laughter*, 26.

35 Patterson quoted in Bryan, *100 Years*, 304. One of the things women did not do at the National Geographic Society was get pregnant, or if they did, let their pregnancy show. Women employees who got pregnant had to resign after three months. Patterson was fortunate enough to get her job back as legends editor three months after her forced resignation in the late 1960s [Patterson, *Lands, Legends, and Laughter*].

36 Bryan, *100 Years*, 33. Scidmore had established herself as a travel writer and photographer in the 1880s with her articles based on trips throughout the northwest coast of

Scidmore overlapped with Ida Tarbell, who was listed as associate editor from February 1901 through February 1905. This may have been a mostly honorary position for Tarbell, a well-known writer with *McClure's*. McClure's company started its brief period of publishing the *National Geographic Magazine* two months before Tarbell appeared on the *Geographic*'s masthead. At the time, Tarbell was working on her momentous Standard Oil articles for *McClure's*, and in her autobiography, the only association with the Geographic she mentions is a purely social one with the Hubbards and the Bells.[37]

Women's occupations at the National Geographic Society, as was the case almost everywhere, were divided by class. Unmarried college graduates, many of them debutantes, might work at the secretary's desk, or perhaps in the photo research department, or on the school bulletin staff. "That a woman might wish to be a writer or editor, as well as a wife and possibly even a mother, was regarded as unladylike, pushy, vulgar, unnatural, and worst of all, non-Geographic," Patterson noted.[38] Women on the cafeteria or cleaning staff were in a different gender category by dint of their class; the Geographic apparently allowed for the possibility that these women worked to support their household. And, as mentioned earlier, the cafeteria dining rooms were also divided by gender and occupation. The one woman staff writer just after World War II, Lonnelle Aikman, "tolerated as a nonmember" by the gentleman's club, ate with the library and research staff in the women's dining room rather than with the other staff writers in their all-male dining room.[39]

As a freelancer, Adams did not have to confront the gender-bound limitations faced by female staff or those who hoped to be on the writing or editing staff of the *Geographic*. Indeed, the *Geographic*'s pages were not closed to women; female travelers and adventurers, including Amelia Earhart, contributed articles and

---

the Americas, from Vancouver to Alaska. She later wrote and photographed extensively about Asia, especially Japan, and was the prime mover behind the planting of Washington, D.C.'s famous cherry trees, which came from Japan. See Sarah Booth Conroy, "She Painted the Town Pink," *Washington Post*, 1 February 1999, A1; http://www.ladnerslanding.com/Newsletters/Vol_2/March_2002.html, accessed 21 April 2006; http://www.explorenorth.com/library/bios/scidmore.html, accessed 21 April 2006.

37 Ida M. Tarbell, *All in a Day's Work: An Autobiography* (Urbana and Chicago: University of Illinois Press, 2003), 181–183.

38 Patterson, *Lands, Legends and Laughter,* 28.

39 Ibid, 28. Aikman was also married, to reporter and writer Duncan Aikman; as Lonnelle Davison, she published her first *National Geographic* article, "Platinum in the World's Work," in 1937. Patterson does not mention where the cafeteria and cleaning staff ate, but it does not appear to be in the same room; Abramson quotes a 1977 *Washington Post* article enumerating separate cafeterias for "men, women, blacks, blue-collar workers," which suggests that black employees, who seem to have mostly comprised the janitorial staff, had their own dining area – or were kept out of the others. (Abramson, *Behind America's Lens*, 148.) Did black men and women eat together? Much of the country maintained segregation between blacks and whites, particularly in the legally codified South. Located between North and South, Washington tended to be a Southern city, especially in terms of race relations.

illustrations, as did wives of exotically located missionaries, diplomats and colonial administrators. Adams may have been turned down for expedition funding, but many of the stories she proposed to the Geographic were accepted. In that respect, the National Geographic Society resembled many other geographic societies at the time, including women to an extent without offering them full membership.

"Through a woman's eyes"[40]: Gender and exploration

Being an explorer was a singular and exotic occupation to begin with; being a female explorer was even more novel. Much scholarship has delved into the highly gendered and racialized activity of exploration, with its attendant contexts of science and empire.[41] Broadly, feminized nature was to be conquered by masculine strength, ingenuity and science; the "children of nature," indigenous peoples outside of Western centers, were to be conquered as well. In the United States, during the period of Harriet Chalmers Adams's career, men such as Theodore Roosevelt, military hero, U.S. president, corporation buster and wild game hunter, and Ernest Thompson Seton, founder of the Boy Scouts, personified the white, upper-class, Anglo-Saxon manly man saved from the fineries of civilization through the practice of roughing it in the wilderness.[42] Exploration, with the social goal of acquiring scientific knowledge and the personal goal of proving one's physical and mental strength, was despite exceptions to the rule, explicitly coded as a white man's game.

And so Harriet Chalmers Adams was repeatedly asked to explain herself. An otherwise feminine person, why did she engage in exploration – not just a masculine pursuit, but the very measure of the utmost masculinity? Upon her return from her first extensive journey in South America, a reporter commented that "few women would have the hardihood to attempt such an undertaking." But Adams sidestepped the question of gender, replying simply that she was just being a freedom-loving Californian. "Why do I do it? Well, for the reason that I want to accomplish something out of the ordinary, to keep up the little record of my own that I have made in South America, you might say, and – oh, I guess mainly because I love the life and liberty of it."[43]

Seven years later she held onto her exceptionalist framework that "what might seem hard for most women is merely play for me." As for what she got out of enduring the hardships of climbing peaks and penetrating jungles, she turned the tables. "Why do men wish to explore new lands? To see and explore new people, to have adventures out of the humdrum of life and to coordinate all these into a

40 "South America Through a Woman's Eyes," *New York Herald*, 5 August 1906.

41 See, for example, Haraway, *Primate Visions*, particularly "Teddy Bear Patriarchy," 26–58; Bloom, *Gender on Ice*; McClintock, *Imperial Leather*; Richard Phillips, *Mapping Men and Empire*..

42 Bloom, *Gender on Ice*, 32, 34.

43 "Tells of Wonderful Travels in Unknown South America," *The* [Sacramento] *Evening Bee*, 1 October 1907.

useful mission for mankind and for their own nation especially. Very likely, this same initiative which impels men impels me."[44]

By 1920, the year that U.S. women got the vote, Adams was able to confront such gender constructions head on.

> There is no reason why a woman cannot go wherever a man goes – and further. If a woman be fond of travel, if she has love of the strange, the mysterious and the lost, there is nothing that will keep her at home – all that it needed for it, as in all other things, is the driving passion and the love. As for being hindered because she is a woman, that sounds rather a stupid notion to me – she has proved that she can go where men go, and that lack of caution (which is a little of every women) will take her out of the tent to discover something else, while her husband is sleeping, covered up to the ears in blankets and fur.[45]

Rather than play up her own exceptionalism, Adams instead recast exploration as a natural activity for women. Without removing "woman" from her "natural" domestic sphere, by her husband's side, Adams rendered the domestic portable (the tent) and conversely placed "exploring woman" in a space no longer necessarily masculine. Indeed, she argued, being a woman could have its advantages. If a woman accompanied a man in adventurous travels in out-of-the-way places, she said, the man would be safer than if he traveled with any number of male companions; "primitive people" were more likely to be accepting of female interlopers than male ones.[46]

"Women often see things about the life and ways of people which a man would not notice," Grosvenor noted after several decades of editing the magazine, proposing that the magazine make a greater effort to cultivate female writers.[47] Somewhat disingenuously, he suggested that perhaps male staffers were more forward than female staffers when it came to requesting assignments, somehow failing to see the segregation, employment restrictions and promotional ceilings at the Geographic as impediments. It would take another couple of decades before the magazine developed a more representative sample of writers.

Although Grosvenor eventually expressed a positive value to women's perspectives, arguments regarding women's difference served for centuries to restrict their entrance into male bastions of knowledge and power. Mona Domosh argues that male geographers deliberately shut "women's knowledges" out of the definition of geography. Works by Victorian women explorers and travelers were denied as "geography" even as men's efforts and tales of exploration were accepted. This was the case because first, exploration was implicitly tied to ideas of masculinity, and

44 *Sunday* [Jacksonville] *Times Union*, 10 May 1914.

45 "Mrs. Franklin Adams Says That Women Can Go Anywhere," Charleston *Daily Mail* [reprint from New York source], 15 December 1920.

46 "Saved by Her Long Hair," [newspaper clipping, late 1920s].

47 Letter from GHG to assistant editor J. R. Hildebrande in 1949, quoted in Bryan, *100 Years*, 299.

second, women's work was judged too subjective, and inferior to men's "objective" texts.[48]

In *Imperial Eyes*, Pratt argues that women have indeed offered travel narratives qualitatively different from those created by men. Comparing works by Flora Tristan and Maria Graham with those of their male contemporaries, Pratt finds that the women's narratives follow neither the Humboltian anti-conquest narrative – rooted in science and imbued with Romanticism – nor the conquest narrative of the "capitalist vanguard." Pratt calls these early-nineteenth-century women "exploratrices sociales," thus emphasizing a certain female focus on interpersonal relations that comes through in these authors' works. "While the vanguardists tend to emplot their accounts as quests for achievement fueled by fantasies of transformation and dominance, the exploratreses emplot quests for self-realization and fantasies of social harmony."[49] Pratt finds that Mary Kingsley, who traveled in West Africa in the late nineteenth century, also defied the standard forms of travel narrative. Commenting that "the masculine heroic discourse of discovery is not readily available to women," Pratt argues that Kingsley undermined the masculine narrative through comic irony and expression of pleasure through play rather than through aesthetics. Kingsley presented a voice "that asserts its own kind of mastery even as it denies domination and parodies power."[50]

Harriet Chalmers Adams was introduced to Kingsley's "book about her travels in Africa" in Valparaiso, Chile, in 1904.[51] Although it is not clear if Adams ever actually read the book, her own writing displays strong parallels with Kingsley's. Both Alison Blunt and Sara Mills describe Kingsley as an anti-colonial imperialist, whose free-trade political beliefs were not incompatible with a desire for more equitable relations between her home continent and the subordinated countries in which she traveled. Nor were her beliefs incompatible with a sympathy for the peoples of West Africa.[52]

Adams had similar leanings regarding Latin America, and actively supported trade between her home country and her favored travel destinations. Like Kingsley, Adams often presented her adventures in humorous hues. In her first *Geographic* article, on Paramaribo, in Suriname, she remarked that the market near her hotel was "pleasantly situated near the city's sewer."[53] Describing balsa boats on Lake Titicaca

48 Mona Domosh, "Toward a Feminist History of Geography," *Transactions of the IBG* NS 16 (1991): 95–104 and "Beyond the Frontiers of Geographical Knowledge," *Transactions of the IBG* NS 16 (1991): 488–490. See also Alison Blunt, *Travel, Gender and Imperialism.*

49 Pratt, *Imperial Eyes*, 168. The term "capitalist vanguard" is Pratt's.

50 Ibid., 213.

51 HCA, diary entry, 31 August 1904. The book Adams'ss friend may have had in mind is probably either *Travels in West Africa* (1897) or *West African Studies* (1899).

52 Sara Mills, *Discourses of Difference: An Analysis of Women's Travel Writing and Colonization* (London: Routledge, 1991); Blunt, *Travel, Gender and Imperialism.*

53 Adams, "Picturesque Paramaribo," *NGM* 18 (June 1907), 373. In a joking toast that deftly linked feminine domesticity with masculine exploration and science, Adams – the only woman invited to speak at the Philadelphia Geographical Society's banquet honoring

for another article in her early series, Adams noted that aboard the boat, "One is in danger of becoming very wet and very seasick. I decided that the boats are most attractive *when seen from the shore*."[54] Adams tempered a possibly conventional tale of heroic adventure with enough self-effacing comments that men frequently commented on her striking combination of courage and modesty.[55]

Although like the men of their race and class, women travel writers reinforced imperialist relations by ignoring political conflict and by leaving unexamined their own presence in the country, Mills argues, their texts offered "counter-hegemonic voices within colonial discourse" largely "through elements such as humour, self-deprecation, statements of affiliation, and descriptions of relationships, which stress the interpersonal nature of travel relationships."[56] In the pages of *National Geographic*, women writing articles seem to have positioned themselves as wives or daughters more often than men positioned themselves as husbands or sons. Adams was no exception. For example, she incorporated her husband, grandfather, and friends into a 1922 lead article on Chile.[57]

While Adams billed herself as an explorer, a designation echoed by the newspapers who covered her lectures and exploits, her editors at the Geographic differentiated between scientific explorer-investigators and travelers. Peary, Macmillan and Chapman, whom the Society funded, belonged to the former category. Adams was placed in the latter. She may have found the "elusive solenodon" as she rode horseback across Haiti, but she also wrote about her fantasy of being an Andean princess. Riding along winding trail in the Valley of Yucay, Adams wrote that she imagined herself "a Quichua princess carried by my willing slaves down to the beautiful summer palace of my father, the Inca."[58] This image did little to encourage

---

Sir Earnest Shackleton – remarked that "Dr. Wiley says that something kept in cold storage for a year is spoiled. Sir Earnest Shackleton disproves this theory completely." "American Woman Makes a Remarkable Tour of South America," *The New York Herald*, 7 August 1910.

54 Adams, "Some Wonderful Sights in the Andean Highlands" *NGM* 19 (September 1908), 604. Italics hers.

55 *Detroit Athletic Club News* (March 1922), in reviewing her lecture to their organization noted: "She was very modest in mentioning her own career as an explorer, but the splendid movies and pictures she showed well displayed the courage it took on her part to complete her South American wanderings." [SPL] Another new fan enthused to John Henry Kellogg about Adams (who stayed at Kellogg's Battle Creek Sanitarium for a short period in 1922 and again after a severe back injury in 1926–27, and gave scheduled talks during her stays): "She has been keenly alert, remarkably observant and has acquired a fund of rare information. And, moreover, what especially appeals to my admiration, is that she is possessed by exceeding rare modesty." Edward M. Brigham to J. H. Kellogg, 16 November 1922, SPL.

56 Mills, *Discourses of Difference*, 22–23.

57 Adams, "A Longitudinal Journey Through Chile," *NGM* 42 (September 1922): 219–273.

58 Adams, "Some Wonderful Sights in the Andean Highlands," *NGM* 19, September 1908, 612.

readers – including editors – to think of Adams as a serious explorer rather than an adventurous traveler.

Men wrote the bulk of *National Geographic*'s travel articles, and both men and women contributed chatty first-person pieces. This "lively" style had been cultivated since the days of Bell's leadership, when he advised Grosvenor to put "a touch of emotion" into stories and write about animals and people to sustain readers' interest.[59]

The most salient divide between women's and men's writing is not so much its content or style, but as Mills argues, "the way that women's writing is judged and processed."[60] Mills's emphasis on the relevance of the *reception* of women's narratives is important in understanding Harriet Chalmers Adams's popularity. Whether it was a woman or man who told a travel and adventure tale or gave a lecture on commercial opportunities made a difference in how the speaker and the subject were perceived, especially since both exploration and business were strongly coded as masculine pursuits. Seeing and hearing the author at a live lecture increased the importance of gender in the narrative's reception. Although it was clear in most of her articles, even if a reader had bypassed the byline, that the author was a woman, when Harriet Chalmers Adams gave a lecture, her gender – and her gender performance – was obvious. Giving a lecture, her feminine delivery was inescapable.[61] A 1915 review in the *Kansas City Post* assured readers that Adams "is essentially feminine and speaks in an unusually sweet and musical voice."[62]

Indeed, Adams deployed her femininity to successful advantage in her career in a man's field, both within National Geographic and beyond. Lectures were her forte, and she enjoyed being on the lecture circuit. "You could not obtain a more popular, entertaining, or instructive lecturer than Mrs. Harriet Chalmers Adams," Grosvenor extolled to Chicago's Chautauqua Institution. "Her pictures are wonderfully beautiful and she has a splendid delivery, a good voice, and her lectures are full of interesting and witty incidents. I cannot recommend her too highly."[63] Adams pleased her audiences – some as large as 1500 to 2000 people – and her "charm" was an important part of her appeal. A petite 5'2", Adams wore explicitly feminine clothes when she lectured; at a lecture she gave for the Geographical Society of

---

59 AGB to GHG, 4 April 1904, Box 267, Bell Papers.

60 Mills, *Discourses of Difference*, 30.

61 One form of reception that Mills discusses was hostile disbelief; people thought that women with wild stories were either lying or exaggerating. Although Mills' period of study, from 1860 to 1930, encompasses most of Adams's travel career, I found no evidence that anyone ever accused Adams of lying or exaggeration. Indeed, articles and letters – and the editorial staff of *National Geographic* – frequently remarked on the truthfulness, accuracy and reliability of Adams's statements.

62 *Kansas City Post*, 25 February 1915. SPL.

63 GHG to Percy H. Boynton, Secretary of Instruction, Chautauqua Institution, Hyde Park Chicago, 9 January 1911, SPL.

Figure 5.2 "It seemed incredible that the small, charming woman, gowned in deep red velvet with a long train, could have visited such strange places"
(Frances Densmore, a noted ethnologist, recalled of an early Adams lecture)
(Society of Woman Geographers, Washington, D.C.)

Philadelphia, Adams was "gowned in an exquisite evening dress."[64] Articles about her, whether they were in general newspapers or directed to a female audience, gave a picture of Adams as an "intrepid little lady," with "sparkling brown eyes," an "irresistible smile," and "beautiful diction."[65]

Both Adams and her husband received letters from first-time attendees pleasantly surprised by her lecture. "About the purely intellectual aspects of the lecturer, there can be no cavil," a Columbia University professor wrote to Franklin Adams, noting that "the attention of her auditors was somewhat distracted owing to their inability to select between the personal charm of the lecturer and the beauty of the pictures."[66] Even Frances Densmore, who became a noted ethnologist and recorder of American Indian music, was struck by the apparent cultural mismatch between the teller and the tale. Regarding one of Adams's 1911 *National Geographic* lectures, Densmore recalled, "It seemed incredible that the small, charming woman, gowned in deep red velvet with a long train, could have visited such strange places."[67]

Although her work challenged stereotypes of male and female roles and abilities, Adams's personal appearance and bearing did not threaten strongly held notions of gender. "Lovely to look at and with a natural charm and sympathy that radiated from her intelligent and understanding personality," reminisced La Gorce in a memorial tribute, "Harriet Chalmers Adams possessed the rare quality of achieving her difficult objective together with the gift of making friends and establishing confidence in her work among peoples in every walk of life, be they rulers of enlightened nations or simple natives in primitive jungles."[68]

By presenting herself as "thoroughly feminine,"[69] Adams found an acceptance that she most likely would not have enjoyed had she presented herself as embodying masculine-coded characteristics such as toughness, pride, and wearing knickers. Adams's femininity was unthreatening on two levels. First, the idea of femininity encompasses accommodation and acquiescence, set in opposition to masculine aggression. Second, while a masculine woman would have been out of the bounds of the bifurcated gender system, Adams's gender presentation suggested that being an explorer was only in addition to – not instead of – her being a "normal" feminine woman. That this distinction was important, particularly in the years before the First

---

64 *Philadelphia North American*, quoted in Adams's promotional pamphlet *The Harriet Chalmers Adams Travel Stories*, SPL.

65 *Springfield Illinois State Register*, quoted in *The Harriet Chalmers Adams Travel Stories*; "Far Away Places Call Again and Noted Washington Woman Explorer Sets Sail Again," *The Washington Post*, 4 September 1933; Wharton, "Woman Turns Geographer"; "Harriet C. Adams in Illustrated Narrative," *San Jose Daily Mercury*, 20 January 1910; all SPL.

66 William R. Shepard, Faculty of Political Science, Columbia University, to Franklin Pierce Adams, SPL.

67 Typewritten memorial, 1937, Box 1, SWG Records.

68 La Gorce, "Harriet Chalmers Adams, Geographer – Writer – Friend," Society of Woman Geographers memorial bulletin, SWG Records.

69 *The* [New York] *Globe and Commercial Advertiser*, 16 November 1908, SPL.

World War, can be seen in reporters' comments on Adams. One New York newspaper declared that it would be "unfair to Harriet Chalmers Adams to call her a women explorer. That makes one think of a pretzel-faced person, dressed in leather clothes and odd knickerbockers, standing, staff in hand on top of a high peak."[70] Kate Davis notes that the reporter's description of a woman explorer "was a nearly perfect portrait of the most well-known photograph of Annie Smith Peck, famed U.S. mountain climber who managed to alienate nearly everyone she met."[71]

Covering Adams's feminine interests was a way to reassure readers of the "normalness" and approachability of their subject, and like Adams's personal appearance, the contradictions between "feminine" and "explorer" made for good copy. "Daily siesta keeps lines from face, says woman explorer" ran the headline of one article.[72] An Associated Press wire story ostensibly about Adams's invitation and coming journey to Haile Selassie's coronation in Ethiopia focused instead on Adams's travel wardrobe, "ranging from flying suits to evening dresses and adapted to both summer and winter climates of 14 countries."[73] Articles about Adams were often located in a newspaper's designated "women's pages."

Blunt notes that Mary Kingsley was also considered "surprisingly feminine and even more feminine than other women despite – and yet because of – having participated in roles regarded as masculine." The contradiction worked, Blunt argues, in part because of the spatial and temporal differentiations in Kingsley's behavior. Her masculine-coded exploits were out of sight of those who would judge her gender-appropriateness, while at "home," she reasserted her femininity. Because her masculine behavior was not on display, it did not pose a challenge to the "patriarchal constructions of women's subordination" at home.[74] It is not as easy, however, to physically separate Adams's "masculine" and "feminine" performance arenas. "Home" for Adams, the California near-frontier, *was* the place where she learned the masculine pursuits of riding, shooting, and exploring. Her American audience, familiar with a valorized narrative

70 Ibid.

71 Davis, "Forgotten Life," n. 37. Peck, who climbed the Matterhorn in leather knickers in 1895, was 60 when she climbed in 1908 what she believed to be the highest mountain in the Western Hemisphere, Mt. Huascaran in Peru, and claimed that she had achieved a higher altitude than any other man or woman. Her achievement completely ruffled Hiram Bingham, who discovered the ruins of Macchu Pichu during what boiled down to a race against Peck to climb a newly-anointed "highest mountain," Coropuna. See Alfred M. Bingham, "Raiders of the Lost City," *American Heritage* 38 (July–August 1987): 54–64.

72 *New York Telegram*, 8 February 1921, SPL.

73 Sue McNamara, "Woman, Abyssinia Bound, Wears Walk-Ride-Fly Suit," n.d. [1930], SPL. "'In out of the way places one has to consider the question of doing laundry,' says Mrs. Adams. 'Rajah silk and lace can be washed out in a bowl, pressed with an iron and come out looking fresh and dainty for next day's wear.'" McNamara seems to have had the Harriet Chalmers Adams beat for the Associated Press; she wrote several articles on Adams or the Society of Women Geographers over a number of years.

74 Blunt, *Travel, Gender and Imperialism*, 161–2.

of white frontierswomen as hearty civilizing forces domesticating America's wilds, would have understood this.

Gender definitions vary by place and time, by class, and race, and age. The span of Adams's career, 1904–1937, was a period of major changes for women in the United States. Although the most prominent political change was the passage of the 19th Amendment in 1920, giving women suffrage, there were several other shifts as well. Women's clothes became more amenable to physical activity, as shapes grew looser and hemlines rose decisively. World War I sent more women into paid employment, foreshadowing the better-known movement of women into the workforce during World War II. The 1920s also saw more women in college and graduate schools than at any time previously.

Adams's own take on feminism grew during these years as well. In the years before the U.S. entered the world war, when reporters asked for her stance on women's suffrage, she described herself as uninterested in the issue, being too busy with her travels and speaking tours to devote much thought to the subject.[75] Of course, what she told reporters was not necessarily what she actually believed, but it would have been how she wanted to be perceived, and her stance was consistent with a general lack of consideration for women as a class. She was simply doing what she wanted, and if she was one of the few, than so be it.

The First World War may have been a turning point in terms of Adams's gender consciousness. Although she presented herself as an exception by being the first American woman war correspondent allowed on the French front, in her 1917 *National Geographic* article, "In French Lorraine," Adams subsumed her special position to that of the women she met. There women literally held the fort, working in the factories and filling what had been, until then, men's positions. While she had discussed women of different groups and classes in her earlier *Geographic* articles, and had written about women of the "other Americas" for the *Ladies Home Journal* in 1916,[76] this was her first *Geographic* story to be so sharply focused on women.

By the 1920s, after the war and her 1919–1920 travels in South America, Adams had become "more consistently focused on women and women's issues in the professional arena,"[77] and she began advocating for forging connections specifically between North and South American women. In her lectures on the status of women in South America, she contended that women of the professional classes there were

75 *Women's National Daily*, 11 January 1911, SPL. According to Blunt, Kingsley was vehemently anti-women's suffrage, and also opposed the admission of women to societies such as the Royal Geographical Society.

76 "Women of Other Americas," *Ladies Home Journal*, October 1916, SPL. "There are traveled women who speak French as fluently as their native tongues, import their gowns from Paris, and have a cosmopolitan viewpoint. Women whose creative genius finds expression beyond their own firesides, educators, painters, musicians, women whose mental horizon is bordered by the walls of their sheltered courtyards. Illiterate peasant women who toil all their days, wild forest creatures who wear tree bark skirts and monkey teeth necklaces. I met them all."

77 Davis, "Forgotten Life," 51.

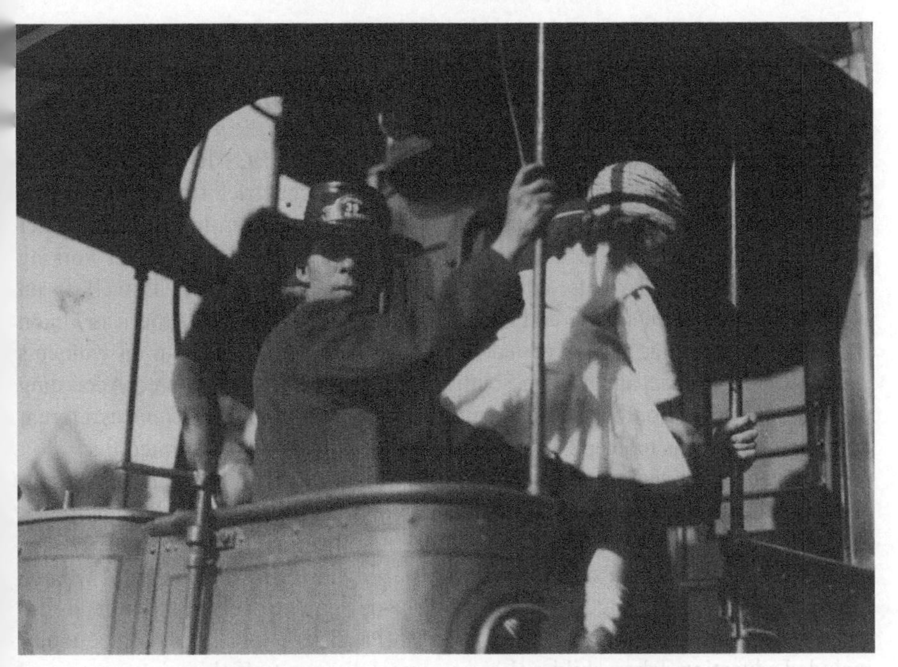

Figure 5.3 "A Woman Street-Car Conductor of Valparaiso"
Harriet Chalmers Adams developed her feminism through her Pan American activities
(Harriet Chalmers Adams/*National Geographic* Image Collection)

more politically advanced than their U.S. peers. "The trouble with us is that we are apt to think that all the women of South America are all little senoritas with black fans and mantillas," she told one New York newspaper in 1921. "Too many do not realize that even twenty years ago the South American woman was active outside her home. What we must do is study her and create a closer bond between her and us," Adams said.[78]

Adams envisioned a Pan American maternal feminism, with women of both continents working together on issues traditionally deemed women's terrain: family, children, and housekeeping. "The great basis for harmony and unity between the two continents, I believe, is to be found in the reforms in the home life and on the basis of mother and child welfare," she told a Midwestern newspaper. "The women of America have the great opportunity of serving as the uniting factor between the two lands, for hygiene and child welfare are the real means of unity rather than politics."[79] Adams's ideas were no doubt influenced by the activism of pioneering

78 *New York City Telegram*, 8 February 1921, SPL.
79 *Battle Creek Moon-Journal*, 13 September 1922, SPL.

women physicians in Argentina, who had spurred national and Pan American conferences on child welfare congress beginning in 1913.[80]

The same woman who ten years earlier had brushed off the issue of U.S. women's suffrage and commented that Latin American women's lack of interest in politics "is not altogether to be regretted perhaps," proceeded in 1921 to comment on the latest suffrage conditions in several Latin American countries from municipal to national levels.[81] Davis suggests that Adams came to feminism in part through her Pan Americanist activities, which introduced her to Latin American women working toward women's civil rights, education and career opportunities, as well as on women's and children's labor and health issues. The monthly Pan American Union *Bulletin*, which Frank Adams edited for several years, had a section on women's issues to which Harriet and leading South American feminists contributed. According to Davis, leaders of South American feminist movements used the *Bulletin* as a forum for debate as well as to keep North American Pan Americanists informed.[82]

The Pan American women's movement was one part of the early-twentieth-century international liberal feminist network,[83] and such famed Anglo feminists such as Carrie Chapman Catt, Jane Addams and Emmeline Pankhurst were involved in the Pan American Women's Conference in 1922, organized by the League of Women Voters. Adams served as a delegate at the three-day conference in Baltimore, as did Brazilian feminist Berta Lutz, and Ecuadorian Pastoriza Flores. Conference topics included education, labor, child welfare, venereal disease, trafficking in women, and marriage laws.[84] The conference helped launch the Inter-American Commission of Women that for the next decade pushed for women's political, economic and social rights at international and national levels.[85]

Adams also became more involved with other North American women proud of their achievements in the non-domestic world. In 1925 she helped form the Society of Woman Geographers (SWG), and was elected its first president. The SWG was

---

80 Donna J. Guy, "The Politics of Pan American Cooperation: Maternalist Feminism and the Child Rights Movement, 1913–1960," *Gender & History* 10 (November 1998): 449–469.

81 "Woman, Caught in Storm of Mountain, Saves Life by Sleeping with Llamas," *The Woman's National Daily*, 11 January 1911; *New York City Telegram*, 8 February 1921, SPL.

82 Davis, "Forgotten Life," 43.

83 Christine Ehrick, "*Madrinas* and Missionaries: Uruguay and the Pan American Women's Movement," *Gender and History* 10 (November 1998): 449–469.

84 *Isabel Keith Macdermott, A Significant Pan American Conference (Washington, D.C.: Government Printing Office, 1922). Digital book, Harvard University Library Page Delivery Service (http://ocp.hul.harvard.edu/ww/outsidelink.html/http://nrs.harvard.edu/urn-3:FHCL:479113, accessed 30 July 2006).* Emmeline Pankhurst represented the National Council of Canada for Combatting Venereal Diseases, and received a "veritable ovation" (Macdermott, 7).

85 See K. Lynn Stoner, "In Four Languages But with One Voice: Division and Solidarity within Pan American Feminism, 1923–1933," in David Sheinin, ed., *Beyond the Ideal: Pan Americanism in Inter-American Affairs* (Westport, Connecticut: Praeger, 2000): 79–94.

the brainchild of Blair Niles, Marguerite Harrison, Gertrude Shelby, and Gertrude Emerson Sen, explorers and adventurers, who felt that "women of the exploring species"[86] should have a place of their own, especially since they were shut out of the New York-based Explorers Club. Although the Explorers Club held teas in honor of women explorers, it would not bend its men-only rules and tradition to invite them to membership, nor would it do so until 1981. "The men, you know, have had their hidebound exclusive explorers' and adventurers' clubs for years and years," Adams told a newspaper reporter, "but they have always been so afraid that some mere woman might penetrate their sanctums of discussion that they don't even permit women in their clubhouses, much less allow them to attend any meetings for discussions that might be mutually helpful."[87]

The Society of Woman Geographers defined "geography" as broadly as did the National Geographic Society, covering world travels as well as scientific field research in social and physical studies. Members met for presentations of their experiences or research as well as for social events. Harriet Chalmers Adams devoted herself to the nascent organization; it was "her child."[88] Branching out from her international work with the Pan American Union, Adams aimed to make the SWG a truly international organization, and she pulled in corresponding members from 39 countries.[89]

Probably because of Adams's leadership, the SWG had branches in both New York City and Washington, D.C., the Adamses' homebase while Frank worked for the Pan American Union.[90] During her eight-year presidency and her honorary presidency following the couple's move to Europe, Adams steered the Society into an international organization in membership as well as scope. "The contacts between specialists all over the world are already proving of value, and we are trying to be generous with one another, and helpful to one another, believing there is room for more than one specialist in any field," Adams told reporters. "We hope in time to render service to young women who have proven their geographical worth and to supply old geographers with understudies where there is danger of their 'going on' without their life being recorded."[91]

Adams's residence in the nation's capital also kept her in regular contact with the leaders at National Geographic, where she was a regular in the Society's lecture series. But much of the time, Adams traveled. The Adamses spent six months

86 Newspaper clipping [title missing], 1 February 1928, SPL.

87 [Newspaper name illegible], 1 February 1928, SPL.

88 Mary Vaux Walcott to Franklin Adams, 12 August 1937, SWG Records.

89 Frances Carpenter Huntington, manuscript for the 25th anniversary of the SWG, 1950, SWG Records.

90 The New York group was more social than the "more scientifically minded" Washington group, according to a Washington-based member. "Perhaps we have so many interesting things transpiring in Washington that we do not need the social angle to keep us stirred up to interest in the Society." Unclear, possibly Mary Vaux Walcott, to HCA, 26 May 1937, SWG Records.

91 Long Beach [California] *Press-Telegram*, 4 April 1928, SPL.

Figure 5.4 **"Ex-head-hunting country, Philippines**
My Little Sister, Ling-Nayo, the prettiest girl in Malay-land." This photograph was used in Harriet Chalmers Adams's lectures, but Adams's travels in the Philippines did not make it into *National Geographic Magazine*
(Society of Woman Geographers, Washington, D.C.)

traveling around Asia and the Pacific – including the Philippines, Japan, China (including Manchuria and the Gobi Desert), Mongolia, Siberia, Sumatra and Borneo. The trip allowed Adams to develop her theory that Native Americans derived their ancestry from Asia, based on not only physical resemblance but what Adams saw as signs of cultural and linguistic similarities. This journey netted no articles for the *Geographic*, although Adams developed a lecture tour and published articles on the Philippines for *The World's Work* and *World Outlook*.[92] After the First World War,

92 Adams, "Uncle Sam's White Magic," *World Outlook* (April 1915); "Snapshots of Philippine America," *The World's Work* (May 1914): 31–42. Did Adams propose any articles from these travels to the *National Geographic*? Did the *National Geographic* editors find Adams's theory unscientific? Did they think her too unschooled to present theories? Did they see her as a Latin American specialist out of her element in Asia? The few Asia articles published from 1913 to 1916 included those by established experts: Dean Worcester, Secretary of the Interior of the Philippine Islands, 1901–1913, "The Non-Christian Peoples of the Philippine Islands," *NGM* 24 (November 1913); John Claude White, a former political officer in charge of Bhutan, "Castles in the Air: Experiences and Journeys in Unknown Bhutan," *NGM* 25 (April 1914); and Eliza R. Scidmore, "Young Japan," *NGM* 26 (July 1914). In 1913, the magazine also ran a traveler's tale of Burma by Charles H. Burnett and a pair of articles on

Adams resumed traveling, revisiting Central and South America, journeying around the French, Spanish and Italian colonies of northern Africa, and spending much of the 1930s living in Europe. Sadly, most of the Adamses' papers, photographs and films were destroyed in a Washington, D.C. apartment building basement flood during their final overseas sojourn. Adams had worried about leaky pipes in the basement and had the building manager check more than once to make sure that everything was fine.[93] But she may never have learned what happened. Harriet Chalmers Adams died in Nice of kidney disease in 1937.

Propaganda of good will: Harriet Chalmers Adams, Pan Americanist

Harriet Chalmers Adams was more than an explorer, but used her experience, observations and knowledge about Latin America to advocate better relations between North and South America. Throughout her career, Adams was adamant that U.S. citizens needed to know more about their "neighbors to the south." Our neighbors know much about us, she argued; it is shameful that we don't know anything about them. "I feel that the lack of harmony and friendship between North and South America has been our fault rather than theirs," Adams told a Michigan reporter, "for we have treated them in a lofty and superior way, and have little understood or appreciated the contributions they have made and are capable of making to civilization."[94] To facilitate knowledge of South America, she advocated teaching Spanish in U.S. schools.

Adams greatly admired the Spanish-influenced culture of the middle classes, and pointed out its virtues on several occasions. Not only should North Americans learn *of* the South Americans, she suggested, but they should learn *from* them as well.

> The Latin-Americans, contrary to opinion, may teach us many things. One is the civic cleanliness. A mosquito or a fly or any other pest is unknown in Rio de Janeiro. … But more important still, is the sweeter family ties and the respect of southern peoples for old people. We could learn much there. We have drifted too far away from the family relationship and we "shelve" our old folks, not regarding with true respect their often saner judgment.

Adams also indicted the North American tendency of "living too fast" and working long stretches of time without a rest, calling the Latin American habit of an afternoon siesta "the art of conserving their energy."[95]

The Adamses had a long connection to Latin America, sharing interests in California's history and honeymooning "by motor" in Mexico. For Frank, his relationship with the "other Americas" became his lifework. From 1907, when he was hired as chief clerk, to his retirement as counselor in 1934, Franklin Adams worked at

---

religion in India. Much of the issues during this period were occupied by articles on Europe, followed by Latin America, the U.S. and Africa.

93 HCA to "Lady," 3 August 1935, SWG Records.

94 *Battle Creek Moon-Journal*, 13 September 1922, SPL.

95 *San Francisco Examiner*, Oakland edition, 15 May 1921, SPL.

the Pan American Union (PAU), an organization that later became the Organization of American States. He edited the Pan American *Bulletin* for many years and also organized performances and international tours of musical groups from throughout the Americas.[96] Harriet devoted much of her career to Pan Americanism as well. She was a keynote speaker at the first Pan American Commercial Conference in February 1911,[97] wrote articles for the *Bulletin*, and was lauded for contributing "considerably to the entente that exists between the Latin American republics and the United States."[98]

Originally the Commercial Bureau of the American Republics, the PAU developed out of the first Pan American Conference in 1890. The stated goal of that U.S.-initiated conference, and of the Pan American movement and organization that followed, was to promote cooperation among Western Hemisphere countries, primarily by standardizing trade and finance regulations. It was clear to many, though, especially Latin Americans, that the conference had more to do with the U.S. concerns that the British and other Europeans were expanding their Latin American trade at the expense of U.S. companies and that the organization was a mechanism for maintaining U.S. economic, political and ideological primacy in the region.[99] As a movement, Pan Americanism was a directed component of liberal-developmentalism, conflating the rhetorics of open markets and progressive uplift. Although Latin American political and business leaders were active participants in the Pan American Union and its conferences, for many Latin Americans, and some North Americans as well, Pan Americanism was "the friendly face of U.S. dominance in the hemisphere."[100]

At least there *was* a friendly face. Under John Barrett, the PAU director-general who hired Franklin Adams, the organization's mission was "to promote peace and stability through freer trade practices, and thus lessen the likelihood that the United States would intervene with force in Latin America. This was easier said than done."[101] In 1904, U.S. president Theodore Roosevelt's Corollary to the Monroe Doctrine essentially declared the United States' right of police power in the Western

96 John Barrett manuscript note for Harriet Chalmers Adams memorial, Part I: Container 1, SWG Records; Franklin P. Adams obituary, *The New York Times*, 11 October 1940, 21.

97 Program, Pan American Commercial Conference, 13–17 February 1911, SPL.

98 Julius Moritzen, *The Peace Movement of America* (New York: GP Putnam Sons, 1912).

99 Arnold Roller, "Pan American Union?," *The Nation* 126 (18 January 1928): 78–80; J. F. Normano, *The Struggle for South America: Economy and Ideology* (Boston: Houghton Mifflin Company, 1931); Arthur P. Whitaker, *The Western Hemisphere Idea: Its Rise and Decline* (Ithaca: Cornell University Press, 1954); David Sheinin, "Rethinking Pan Americanism: An Introduction," in Sheinin, *Beyond the Ideal*, 1–8; Joseph Smith, "The First Conference of American States (1889–1890) and the Early Pan American Policy of the United States," in Sheinin, *Beyond the Ideal*, 19–32.

100 Sheinin, "Rethinking Pan Americanism," 1.

101 Salvatore Prisco, "John Barrett and Collective Approaches to United States Foreign Policy in Latin America, 1907–20," *Diplomacy & Statecraft* 14 (September 2003), 59.

Hemisphere, and the U.S. took over the finances of the Dominican Republic. The U.S. military intervened in Cuba from 1906 through 1909, and in Nicaragua in 1909 and 1912, maintaining control of the Nicaraguan government until 1925 (only to be followed by another invasion in 1926). In 1914, the U.S. invaded Mexico during its revolution; in 1915, the U.S. invaded Haiti, occupying it until 1934. Its Hispaniola neighbor, the Dominican Republic, saw a U.S. invasion and takeover in 1916, the same year the U.S. invaded Mexico again.[102]

Still, United States Americans working with and in the Pan American Union believed in their ideal of peace through free trade and cultural exchange, and did not find it contradictory to hold that Latin America's free trade should be freer with the United States than with Europe. John Barrett advocated multilateral decisions, toll-free access for Latin American nations through the Panama Canal, and a Pan American-mediated solution to the crisis in Mexico, all dismissed by the U.S. administrations under Taft and Wilson.[103] Barrett's successor, Leo Stanton Rowe, who headed the PAU from 1920 to 1946, also opposed U.S. military intervention and believed that U.S. corporate interference in Latin American politics was "a great source of mistrust between Latin America and the United States." While he held that monopolies were bad for business (other than their own) and diplomatic relations, Rowe believed that economic development in Latin America would drive positive social change. His was a progressive paternalistic approach, working off an acceptance of uneven development between North and South: U.S. business would benefit from capital investments in Latin America, and Latin America would benefit from U.S. funds for infrastructure and industrialization.[104]

Harriet Chalmers Adams also advocated for more and better trade relations between the United States and Latin America, but she liked to stress that Latin Americans were not the only ones with something to learn. Among other Pan Americanists, Adams believed that U.S. ignorance of South America was a leading cause of the United States' lagging behind Europe in trade with that region. To encourage investment in and trade with Latin America, it was important that U.S. actors not only know more about, but feel positively towards, the region. Her 1910 *National Geographic* article on the railway built through the Andes, for example, explicitly links knowledge of South America with a struggle for U.S. trade hegemony in the region.

> With the opening of the Panama Canal we North Americans will have a golden opportunity to win from Germany and England the trade which is ours by the right of contiguity. Through gross lack of understanding of our Southern neighbors, we have lagged behind in the commercial race. ... We owe better acquaintance to our Latin sisters. We owe

102 Cyrus Veeser, *A World Safe for Capitalism: Dollar Diplomacy and America's Rise to Global Power* (New York: Columbia University Press, 2002); Prisco, "John Barrett," 59; Benjamin Keen and Keith Haynes, *A History of Latin America* (7th Ed.), (Boston: Houghton Mifflin 2004): 531–2.

103 Prisco, "John Barrett," 59–60.

104 David Barton Castle, "Leo Stanton Rowe and the Meaning of Pan Americanism," in Sheinin, *Beyond the Ideal*, 33–44.

> commercial advancement to ourselves. … We should take first rank in the near future on South American trade.[105]

The connection Adams makes for the *Geographic*'s readers between "pure" knowledge and pure profit is explicit. The *Bulletin of the International Bureau of the American Republics* (an earlier name of the Pan American Union) called attention to another of Adams's *National Geographic* articles in that series, noting, "If other parts of Latin-America always received the same sympathetic treatment from cultural observers that Mrs. Adams gives to Peru, our travel, study, and commerce would be immensely stimulated in that direction."[106] In that article, the lead story for *National Geographic*'s October 1908 issue, Adams noted that Germans got the trade because they would go to out-of-the-way places like Cuzco, which was not on a train line at the time, and "study the needs and tastes of the people. If the descendants of the Incas yearn to wear pea-green and royal purple, the Kaiser's commercial travelers plan that they may."[107]

Adams's articles for *National Geographic* emphasized commercial and cultural geography over a heroic narrative, although she could spin a heroic adventure tale with aplomb. Davis suggests that Adams's more adventurous explorations never made their way to publication in *National Geographic* because of its adherence to gender definitions in which wild adventure was inappropriate for those expected to embody the ideal of white women's civility.[108] But it is also likely that Adams gave the *Geographic* the sort of fact-filled, descriptive article she thought they wanted, anticipating that the magazine's readership would appreciate the economic information and advice.

Adams spoke directly to North American businessmen on a number of occasions. "She has been able to not only gain the friendship of Latin Americans, but she has been exceptionally successful in bringing the commercial chances before American businessmen," noted one contemporary observer.[109] Her lecture to the 3,000 delegates of the Associated Advertising Clubs of America convention on "advertising and export trade in relation in Latin America" in 1911 received "the greatest ovation of

---

105 Adams, "The First Transandine Railroad from Buenos Aires to Valparaiso," *NGM* 21 (May 1910), 417.

106 *Bulletin of the International Bureau of the American Republics*, November 1908, SPL: "No more charming and illustrative of certain phases of life and reflections of history has ever been writtenthan that in the 'National Geographic Magazine' (Washington) for October, 1908, by Harriet Chalmers Adams on Cuzco, America's ancient mecca. Mrs. Adams is inspired for her work by a sincere love of the beautiful, by an innate appreciation of the picturesque, and by a humane philosophy which enables her to look at both the past and the present in true proportion." Given Frank's employment at the International Bureau of the American Republics at the time, possibly already at the *Bulletin*, chances were good that the writer knew Harriet.

107 Adams, "Cuzco, America's Ancient Mecca," *NGM* 19 (October 1908), 687.

108 Davis, "Forgotten Life."

109 Moritzen, *Peace Movement.*

the three days of the meeting."[110] There she reiterated the importance of knowledge of Latin America for successful U.S. commerce in that region. Ignorance is bad for business, she declared, citing the lesson of a U.S. mail order company that printed a catalog in Spanish for circulation in Haiti and Brazil.[111]

World War I shifted the balance of commerce with South America; the United States benefited from the war's interruption of European- South American trade. On her return from her 1919–1920 travels in South America, "where she was sent by the United States government to study commercial conditions," Adams continued to advocate for U.S. commercial involvement in South America, speaking "before prominent business men's organizations."[112] "Mrs. Adams thinks that American men of business and governmental US are 'asleep at the switch' to let other nations from afar come into the southern continent that rightfully belongs to us," reported the San Francisco *Examiner* in 1921. "I fully believe that the two continents of the western hemisphere are going to need one another immeasurably in the years to come," Adams told the newspaper. "Our greatest danger now is apathy, in letting go of the great unifying principles we developed during the great struggle and in lack of interest in protecting this country in the future by propaganda of good will where it will do us the most good, that is, in the countries of South America."[113]

Paths of progress

Like many affiliated with National Geographic, and with Pan Americanism, Harriet Chalmers Adams operated within a liberal-developmentalist ideology that advocated expanding trade, stressing the natural and desirable trajectory of economic development through increasing involvement with a global capitalist economy, and approving of political imperialism only as long as it focused on moving the process of progress along. In the post–1898 United States, that narrative found a home base in the Philippines and as noted earlier, a forum at *National Geographic*.

110 "3,000 Advertising Delegates Cheer Woman Traveler," *The Philadelphia Times*, 3 August 1911, SPL. Adams was one of the first women in the history of the organization to take a formal part in the convention. See Helen Mar Shaw-Thompson, "Women Favor Advertising," *The New York Times*, 4 August 1911, 3.

111 Davis, "Forgotten Life." Adams also explained to her audience the preferred forms of advertising in South American countries, and pointed out that the Germans fared so well commercially in South America because, unlike the North Americans who tried to sell the South Americans items they did not want, the Germans determined first what it was that their market wanted.

112 "U.S. Is Asleep at the Switch, Says Observer," *San Francisco Examiner*, 15 May 1921. While the suggestion that Adams was sent to South America by the U.S. government to study commercial conditions is intriguing, Adams herself made no mention of such employment, nor do other newspaper articles. She may have gone with Frank under the aegis of the Pan American Union, which the newspaper reporter may have believed to have been a governmental organization.

113 Ibid.

Although the *Geographic* ran many articles on the Philippines, none were written by Adams, despite her travels there and subsequent lecture tour. Adams's Philippine experiences and observations found printed homes in two other publications, *The World's Work*, the monthly current affairs magazine published by Doubleday, Page & Company, and the short-lived Methodist mission magazine *World Outlook*.[114] In the *World Outlook* piece, Adams cheerfully declared that "Uncle Sam's white magic … has turned a backward people from tribal quarrels and primitive ways of living into the path of unity and modern progress."[115] Specifying ethnographic details in *The World's Work*, Adams discussed traveling from Ifugao, where the girls wear only a wrapped "strip of cloth … from waist to knee," to Cervantes, a "Christian Filipino village in Lepanto Province," where the girls are "a good step up in evolution."[116]

Contrasting, if vaguely, French rule in Algeria with that in Morocco, Adams noted in a 1925 *National Geographic* article that the French learned from their mistakes in Algeria and Tunisia and so "have made a steady advance in pacification, unification, and progress" in Morocco. "Abreast with the troops, in greater part native, but officered by Frenchmen, have marched the roads, railroads, schools, hospitals, and agricultural bureaus."[117] Building infrastructure was the civilized and peaceful way of asserting control over a subject people. Such activity could be seen as altruistic, although it was also, of course, the means by which the area's riches could be more easily transported and managed.

In an expressly political piece for *The American Review of Reviews*, in which Adams interviewed Spain's military dictator, General Manuel Primo de Rivera, less than two years after his coup, Adams reiterated her admiration for the developmentalist approach to imperialism. De Rivera, she wrote, was following in Spanish Morocco "those methods employed with such brilliant success in French Morocco, where roads rather than regiments have won the victory."[118] Although she favored "roads rather than regiments" and economic sway over military aggression, Adams expressed the inevitability of the strong triumphing over the weak. "Not that I personally lack sympathy with the Berbers," she wrote. "North Africa was their land even in those shadowy days before the first Phoenician craft sailed toward the Pillars of Hercules. But like the primitive American peoples, their race is run."[119]

The march of history had a prominent place in Adams's articles for the *Geographic*. Through her eyes, readers could imagine wave after wave of conquering nation or generations of leaders and adventurers. Stone and marble ruins told the tale in Libya's

---

114 *The World's Work* was founded in 1900 by Walter Hynes Page and edited by him until U.S. President Woodrow Wilson appointed him ambassador to Great Britain in 1913. See Robert J. Rusnak, Walter Hynes Page and *The World's Work*, 1900–1913 (Washington: University Press of America, 1982). His son Arthur W. Page replaced him as editor, leaving in 1927 to head public relations for AT&T and became known as the father of PR.

115 Adams, "Uncle Sam's White Magic," *World Outlook* (April 1915), 11–13.

116 Adams, "Snapshots of Philippine America," 34–35.

117 Adams, "Across French and Spanish Morocco." *NGM* 47 (March 1925), 352.

118 Adams, "Truth About Spain," 72.

119 Ibid.

Figure 5.5 "A Bedouin Family Rides to Town on Its Two-Seater, Desert Model, Roadster"
Reflecting on her travels in colonized North Africa in the 1920s, Harriet Chalmers Adams expressed sympathy for its conquered or still fighting peoples. "But like the primitive American peoples, their race is run."
(Harriet Chalmers Adams/*National Geographic* Image Collection)

Cirenaica: "Here the Egyptians followed the Greeks, and the Romans followed the Egyptians; then came the Byzantines."[120] In Rio de Janeiro, Adams speculated on which Portuguese explorer first set foot in "this marvelously beautiful landlocked haven" and took readers to a small city park "where, in 1792, the first martyr of Brazilian liberty,

120 Adams, "Cirenaica, Eastern Wing of Italian Libia." *NGM* 57 (June 1930), 690.

Sublieutenant Joaquin José da Silva Xavier, nicknamed 'Tiradentes,' or Tooth-Puller, was executed." He is sure to be celebrated, Adams wrote, in the upcoming centennial of Brazil's independence, part of "the splendid historical pageant even now being staged."[121]

Taking the long historical view, Adams did not believe that greatness was confined to white people. She never ceased to marvel at the Incas and other pre-European American civilizations. Compared to the solid stone slate walls the Incas built in Cuzco, she wrote, "the Spanish edifices look crude and decayed. There is a strength and dignity in this work of the ancients."[122] She trumpeted the "splendid resistance" of the Araucana, or Mapuche, in Chile to successive waves of Incas, Spanish and Chileans – although the Chileans eventually conquered them. Somewhat sorry to see a once-proud culture become absorbed, Adams found cheer in the biological blending of Mapuche and European Chilean, so that rather than annihilation, "the splendid physique and valorous traits of this native people will not be lost to posterity."[123]

In another in her early series of Andean articles, Adams mused about the cycles of civilizations through history and into the future. Meditating on a crumbling pre-Incan doorway, Adams imagined the massive wall and populous city of which it was once a part, "contemporaneous with the ancient capitals of Egypt and the East. I did not feel as confident of our triumphant modern civilization," she wrote. "'History repeats itself,' the thought came to me; 'civilizations rise and fall.' Which of the mighty edifices now standing in America will testify to our nation's greatness in the centuries to come?"[124] Hers was clearly a civilization on the rise, but she could allow the possibility that the day might come when, like the Berbers and the Native Americans, her race too would be "run."

Her ability to take the long view was part of a general inclination to consider things and people openly. Adams had not found the Andean peoples particularly friendly or engaging, but she read their "sullen, lifeless" qualities as those of a people squashed.[125] "Theirs has been such a heart-rending history," she wrote of the Quichua in Peru. "What blessing has European civilization brought to them which they did not already enjoy? What have they not suffered in the name of the cross which surmounts the hill?"[126] A decade later, in the Philippines, Adams tempered her statement that the Ifugao were less evolved than the "civilized" Filipinos by referring to them in terms of "dignity" and "respect." She noted that Ifugao girls, with their knee-length skirts, "can walk much better than I, with the present fashion

---

121 Adams, "Rio de Janeiro, In the Land of the Lure," *NGM* 38 (September 1920), 185, 191.

122 Adams, "Cuzco, America's Ancient Mecca," 676.

123 Adams, "A Longitudinal Journey Through Chile," 249, 255.

124 Adams, "Some Wonderful Sights in the Andean Highlands" *NGM* 19 (September 1908), 599.

125 Ibid., 618.

126 Adams, "Cuzco, America's Ancient Mecca," 689.

in skirts." Comparing the contents of her bag with that of a "savage … chief," Adams noted that his things were "all well made and useful, all ingenious and known to his people since time immemorial. On the whole his trail equipment was better than mine and I had to acknowledge it."[127]

Adams captured the standard *National Geographic* approach of touting the spread of Western progress while mourning the loss of overshadowed and altered cultures. Like Maynard Owen Williams, Adams wanted to see and capture places before they had been "spoiled" and changed by capitalist intrusions.[128] With the right landscape, though, Adams could easily time travel to an earlier authentic, even prehistoric, time. "[T]hroughout the Andean highlands the traveler feels transported to centuries gone by," Adams reported in 1908. At Lake Titicaca, "as I watched a fleet of balsas sailing out the fishing grounds I realized that in the people, crafts and lake itself there is little change since prehistoric days." In 1925, traveling with her sister and several other passengers in a bumpy jitney, Adams "had a feeling of gratitude that we had not come to inland Morocco too late. Here, little changed since prehistoric days, were the Berber people. Here, right in the automobile with us, were natives taking their first ride!"[129] With the recent introduction of planes, trains and automobiles, Adams wrote, the Berbers had suddenly been catapulted from prehistory to the modern world. Despite her grasp of the historical waves of invasion and change, whether in North Africa or South America, Adams was still stuck in the trope of the unchanging primitive.

Like so many others writing for the *Geographic*, Adams had a fascination with racial typing. She was particularly intrigued by what she saw as the resemblances between different Native American tribes and Central Asian or Pacific Island peoples, and speculated that the first Americans arrived by boat across the Pacific. Both early and later articles detail the racial and class characteristics of the people she observed in her travels. Her concept of race was mostly physical but often incorporated layers of cultural history or nationality. She described the people of Morocco, for example, as "a mixed race with a Berber base." Having previously described her Berber fellow-passengers as "swarthy, features rather flat, eyes small and brown," Adams then rescued them from total otherness by asserting that Berbers, if one goes back far enough, are actually white, "allied to a prehistoric race of southwestern Europe." Some "aristocratic types" are "men of white skin and noble mien" while the genetic mix – or to be more historically appropriate, the blood admixture – of the Moors

127 Adams, "Snapshots of Philippine America," 33–34.

128 In "Cuzco, America's Mecca," 687, Adams wrote that "The dreaded time is coming when [the Andean Indians] will forsake their picturesque homespun altogether for gaudy materials 'made in Germany.'" Williams, who generally wrote his captions, explained of a young woman's photograph, '"Made in Germany" is stamped all over the garish dress of this Euphrates Valley maiden.'" (Williams, "Syria: The Land Link of History's Chain," *NGM* 36 (November 1919), 448.

129 Adams, "Across French and Spanish Morocco," 331–332.

includes "Arab, Jewish, Turkish, European. And there is also Negro slave blood from the far south."[130] Rather than including "Negro" in the list, Adams added them in a separate sentence and in so doing, distanced these gene pool contributors even further by rendering them specifically enslaved and originating far away.

Adams's racism was less subtle in 1907 when she explained the multicultural population of Trinidad to *National Geographic* readers. She repeatedly and favorably compared the East Indians who came to Trinidad as indentured servants to the Africans whose ancestors were brought to the island as slaves. East Indians were graceful, strong, and steadfast workers; the women had "appealing eyes" and the men "those all-knowing eyes of the Far East." The "coarse" blacks, on the other hand, were characterized by "idleness" and "indolence." In comparison with the blacks, "these people of an ancient race [the East Indians] stand out in the traveler's remembrance as a more fitting type in lands of such great natural beauty."[131] "Race" here is both physical and cultural; ancient refers to civilization, not bloodlines. Adams's dislike of the Trinidadian blacks allowed her to deny them history, rendering them as people without architecture, with no civilization, as she denied them the acknowledgement she granted the Quichua about the consequences of oppression.[132]

Like Williams, Harriet Chalmers Adams considered herself unjudgmental and open to the world. But like Williams, like the Geographic as a whole and like so many other white Americans of her time, she did not seem to recognize her deep-seated prejudice against blacks as a contradiction and an impediment to such openness. The familiar Other was not the exotic romantic but a figure of contempt. Still, Adams had cordial relations with middle-class African-Americans. Franklin Adams had a warm relationship with John A. Simms III, an African-American with whom he worked for many years at the Pan American Union, and Harriet sent him a brief letter of appreciation after the Adamses left for Europe in 1933.[133]

---

130 Ibid., 332, 345.

131 Adams, "East Indians," 491, 485.

132 She showed the same lack of regard for black Africans in her request for a West African expedition from Kano, in Southern Nigeria, to Jenne and Timbuktu. "I understand … that the peoples found all along this route are a thousand times more interesting than those to the south in the more tropic region, where germs abound," she wrote. "It is a part of the world that should yield much of interest to you, in pictures and in much else, for it is an historic land and the people who live there are not the savages that one finds to the south in Congo country." HCA to GHG, 29 March 1916, NGS Records Library. Perhaps even Grosvenor was put off by her characterization of central Africans as uninteresting savages.

133 HCA to John A. Simms III, Simms Family Papers, Box 89–1, Folder John A. Simms III Correspondence, Manuscript Division, Moorland-Spingarn Research Center, Howard University. Frank Adams'ss letters to Simms in this collection, all after Harriet's death, are open and relaxed, as talking to a confidant and an equal, and detail his grief with her absence.

Through the class doors

Just as a woman could participate in the masculine activity of exploration and return unscathed as "woman," as did Harriet Chalmers Adams, so too could a woman engage in the role of "seeing-man." "Seeing-man" is not only gendered but classed; one looks from above, distanced, possibly possessive, but certainly without fear or consciousness of subjugation. Harriet Chalmers Adams's adventurous spirit was her own, but she was a woman very much of her class. Within the wild California frontier girl was a solid bourgeois colonial New Englander of British heritage. Her background gave her access to leisure, freedom of movement, international social connections, and entry into the elite world of geographical and explorers' societies, and it informed her political and social views. It was her class – in the several connotations of the word – that opened the doors to her membership in the National Geographic Society family, to its magazine pages, its lecture circuit, and its inner circle.

Like so many others in her class, and in her circles, Adams conformed to the prevailing political-economic ideology, depicting conquest and exertions of power as part of a natural turn of events. Although she had mixed feelings about the path that history had taken the indigenous peoples of the Americas, Adams absorbed and perpetrated the liberal-developmentalist model of social evolution through U.S. economic expansion generally favored by *National Geographic*. In an odd way, Adams's great admiration for all things Spanish tempered the American chauvinism so inherent in the liberal-developmentalist ideology. As much as she worked for U.S. trade supremacy in South America over European countries and Japan, Adams's Pan Americanism embraced, it seems, a genuine interest in equalizing relations between North and South America. She identified strongly with South Americans, although her identification was class-bound (Creole/middle/professional) and often attached to gender (female). Her feminism was a cosmopolitan South American feminism, inspired by the women who participated in Pan Americanism from a South American perspective. Her defense of Latin American culture and of the worth of South American civilization, worked as well to question the desirability of North American hegemony in the South.

Like Maynard Owen Williams, Harriet Chalmers Adams can be seen as a good representative of a National Geographic contributor. Her anecdotal, detail-filled travel stories were consistent with the *Geographic*'s liberal-developmentalist model, touting progress while acknowledging nostalgia for "authentic" bygone societies. "The thing to bear in mind is that characters in history were human beings in their time," Adams told her friend and fellow scholar of Latin American history, Irene Wright. "Some conquered kingdoms for Spain, but they also watered geraniums – at least, their mothers, wives and daughters did."[134] Looking back at Adams, at Williams, at Grosvenor and their colleagues and co-contributors at *National Geographic* from the early days of the twenty-first century, it is helpful to

---

134 Irene Wright, recalling HCA's words in her 1937 memorial, SWG Records.

keep her reflection in mind. People of their time, they both reflected the dominant views of their time and place and actively worked to communicate and naturalize their perspectives to hundreds of thousands of readers of the *National Geographic Magazine*.[135]

---

135 Adams languished in historical obscurity for several decades after her death, and was not even mentioned in Bryan's official centennial history of the Geographic. Since then, though, she has appeared as the subject of a small-press young adult biography in 1997, in a 2000 collection of women adventurers published by National Geographic, and from the same year, in a collection of women photographers from the Geographic. These celebrative portraits rescue Adams as an interesting and intrepid person and an inspirational pioneer, but leave her political perspectives and efforts (excepting the founding of the Society of Woman Geographers) out of the picture. See Anema, *Harriet Chalmers Adams*; Michele Slung, *Living With Cannibals and Other Women's Adventures* (Washington, D.C.: National Geographic Adventure Press, 2000); Cathy Newman, *Women Photographers at National Geographic* (Washington, D.C.: National Geographic Society, 2000). Adams is also discussed in Marion Tinling, *Women Into the Unknown: A Sourcebook on Women Explorers and Travelers* (Westport, CT: Greenwood Press, 1989). And she has a Wikipedia entry: http://en.wikipedia.org/wiki/Harriet_Chalmers_Adams (accessed 21 October 2006).

Afterword

The *National Geographic Magazine*'s place in the cultural landscape of the early twentieth-century United States was a cherished one. Lyndon Baines Johnson, the U.S. president from 1963–1969, remarked that "my mother brought me up by putting the Bible in my right hand and the *National Geographic* magazine in my left."[1] Johnson had grown up relatively poor in central Texas, but the magazine's role in his upbringing paralleled that of U.S. senator Leverett Saltonstall, a patrician from Boston. Responding to Johnson, Saltonstall declared, "My family had the *Geographic* and I was brought up with the same thing!"[2] Other influential Americans grew up reading *National Geographic* as well. Howard Stassen – Minnesota governor, U.S. senator, United Nations charter delegate, University of Pennsylvania president and perennial U.S. presidential candidate – grew up on a Minnesota prairie farm where the *National Geographic Magazine* was "the star magazine of the household."[3]

In a speech at a 1948 Philadelphia Geographical Society award ceremony honoring National Geographic's John Oliver La Gorce, Stassen stressed the twin themes of education and American preeminence. "I feel very strongly that the total impact of the National Geographic Society . . . upon the understanding in this country of the geography and the people of the world cannot be underestimated; and that, in fact, as we think of the increased importance of leadership of our United States of America in the world in the months and years ahead, the importance of the work of this Society of the carrying on of its activities, and of geography as a whole, increases."[4]

That *National Geographic* has indeed been significant in forming Americans' knowledge of "the geography and the people of the world" is, of course, a central premise of this book. The Geographic presented an articulation of what it meant to be American in relation to the rest of the world. To be American meant representing and being represented by strength and benevolence in global relations. Being American meant being both technologically and morally advanced. Being American meant being friendly and having a sense of humor, even if some of that humor was at the expense of the different people in other places. Being American meant being white, of

1 Johnson's mother joined the NGS in 1919, when he was 11 years old. Transcript, Melville Bell Grosvenor Oral History Interview I, 28 April 1969, by Joe B. Frantz, pp. 5, 9, LBJ Library. Online: ftp://webstorage2.mcpa.virginia.edu/library/nara/lbj/oralhistory/grosvenor_melville_1969_0428.pdf (Accessed 28 July 2006.)

2 Ibid., 5.

3 Proceedings of the Philadelphia Geographical Society, Philadelphia, 6 December 1948, Box 159, Grosvenor Papers.

4 Ibid. The Henry Greer Bryant Gold Medal that the Philadelphia society gave La Gorce had been established in 1934 and had been granted previously only to Gilbert H. Grosvenor and Isaiah Bowman.

northern or perhaps central European heritage, and Christian, and it often meant being male. It meant having middle-class comforts, including receiving a monthly magazine that "brings to our feet the out of the way corners of the world," as an early National Geographic promotional letter declared.[5]

The Geographic built its reputation and its membership by presenting itself as a reliable source of information about the world. In part, that reliability came from an understanding of geography as a science, an association from the Society's very beginning. Under editor Gilbert H. Grosvenor, the Geographic's reputation was fostered by a proclaimed effort not to dirty the essential truths of people and places with references to contemporary politics. Of course, by framing the world through such narratives as Orientalism and beneficent imperialism, and by omitting coverage of elements deemed unpleasant or reprehensible, the *National Geographic Magazine* communicated its collected knowledge in an expressly political manner.

These days, although the *National Geographic* magazine still remains popular, Americans are perhaps as likely to come across National Geographic on the Internet, or as the producer of a TV special, or through its cable channel. From its peak of 10.9 million in 1981, membership in the Society, with its attendant subscription to the magazine, dropped to about 6.5 million by 2004.[6] Bell was right, people do gravitate toward pictures, and in the twenty-first century Americans are getting their stories and images of the world from television and the Internet. While National Geographic has much more competition in providing information to United States readers about the world around them, it still has its reputation as a trusted source, a brand that people know and can rely on. As the bond rating company Moody's Investors Service said in a 1997 report, "The National Geographic name remains among the most recognizable brand images in the U.S. and internationally, and is likely to remain so even as competition multiplies."[7] Students who preface a statement in class with "I saw something on the National Geographic channel," are asserting that the information they are about to proffer is true and verifiable.

The small international membership of the Society that Maynard Owen Williams wished to foster has turned into a coveted market. The magazine now has foreign editions in 28 languages. The Society also targets American audiences with publications whose names reveal their intended readership: *National Geographic Traveler*, *National Geographic Adventure*, and *National Geographic Kids*. The National Geographic Society has also expanded its educational mission by publishing a classroom magazine, *National Geographic Explorer*, and funding geographic education programs throughout the United States. Its extensive website allows browsers to check out highlights of the latest issue and television offerings, order National Geographic products, generate and download maps, and *subscribe* to the

---

5 MS. of letter, late 1908–early 1909, Box 160, Grosvenor Papers.

6 Poole, *Explorers House*, 293; Mark J. Miller, "To the Ends of the Earth," *Folio*, 1 March 2004, 3.

7 "National Geographic Gets a Facelift," *The Toronto Star*, 21 August 2000.

National Geographic magazine. These days, membership in the Society is presented as an extra benefit to subscribing, and not the other way around.

Perhaps the most dramatic changes in the organization came as a result of the Society's decision in the late 1990s to become, at least in part, a profit-seeking corporation.[8] The National Geographic Society has remained a tax-exempt nonprofit organization, publishing magazines and books and dispensing grants. Spin-off products such as the television channel, website, and collected issues on CD-ROM fall under the profit-seeking, tax-paying National Geographic Ventures.[9] Although press coverage of the changes expressed surprise at their commercial nature,[10] it should be remembered that ever since the first Gilbert Grosvenor took charge in the early twentieth century, both financial health and growth had been priorities.

The Society's corporatization came on the heels of the 1996 retirement of president Gilbert M. Grosvenor, grandson of Gilbert H. Grosvenor. As Grosvenor moved into the largely ceremonial title of chairman, the new regime shrunk the editorial staff, including staff writers and photographers, and sold the Maryland building housing the subscription and customer-service departments. Those jobs are now subcontracted, as are those of the custodial, cafeteria and security staffs.[11] Only the number of executives and advertising positions has increased.[12]

Despite the substantial changes at the organization, National Geographic continues to embrace the long history that established its reputation. The "Flashback" photograph feature initiated in 1995 is archived on the website, allowing Internet readers to look at highlighted images from the magazine's more distant past. And Maynard Owen Williams's May 1923 article, "At the Tomb of Tutankhamen," now functions as a multilayered meta-narrative.[13] The presentation is designed to take us back in time, opening with a black screen like an old-fashioned movie. We are brought back to the original article through a frame evoking the original oak-bordered *National Geographic Magazine* cover, indicating (imprecisely) a February 1923 publication date. With a click, readers are presented with Williams's dramatic and picturesque, though abridged, story of the uncovering of Tut's tomb.

This interactive feature taps into the Geographic's ongoing fascination with ancient Egypt and ties in with its first IMAX film, *Mysteries of Egypt*. Using Williams's original text, it serves as old-fashioned adventure story and allows the Geographic to reassert its authority on the subject; Williams was one of the first correspondents allowed in the inner chamber of the tomb. As with the original article, readers are meant to feel that they are there with Williams (hence the revised

8 Constance L. Hays, "Seeing Green."

9 Poole, *Explorers House*, 303.

10 Hays, "Seeing Green."

11 Poole, *Explorers House*, 301; Paul Farhi, "Mapping Out a Greater Society; National Geographic Is Pushing Beyond Its Yellow Borders Into New Market Territory," *The Washington Post*, 27 January 1997, F12.

12 Poole, *Explorers House*, 306; Miller, "Ends of the Earth."

13 See www.nationalgeographic.com/egypt (accessed October 21, 2006).

February designation, the month of the tomb's uncovering). But we are not simply readers of the original article. This is the twenty-first century and with another click, we can read a snippet of a letter, appearing to have been typed on Luxor Hotel stationery, that Williams wrote to La Gorce, his editor. In the first excerpt, Williams apologizes for the cost of the assignment and assures La Gorce that he will get his money's worth. He mixes his complaints with gratitude. "My legs curse you. But my heart says "Thank you."[14]

New technology facilitates the extra layer of content, but the narrative remains essentially that of the 1920s. The behind-the-scenes peek does not detract from the spirit of adventure, nor does it overtly question the "right" of non-Egyptians to raid a gravesite and control its contents. It is a technologically, though not critically, enhanced *National Geographic* story that assumes the right of European/American dominance of a colonized country.

As Lutz and Collins discuss in their examination of *National Geographic* through the 1980s, the magazine held fast to its formula into the second half of the twentieth century. The smiles that Maynard Owen Williams pursued for the magazine in the 1920s, 1930s, and 1940s – and that he so passionately defended in 1945 – were still in primary demand for *National Geographic* in the 1980s. "'It behooves us to show reality," one editor said, "and nothing is all bad or all good. If [the photographer] didn't find any happy people, I'd tell him to go back and find them.'"[15] While by that time *National Geographic* had begun occasionally discussing and displaying scenes of poverty and environmental destruction, smiling faces and upbeat captions were employed to counterbalance such stories and photos.[16]

Lutz and Collins argue that while *National Geographic* has been more respectful of non-Western peoples than has much other media, the magazine's humanism is also a conservative force. The magazine's soft focus "obscure[s] the American relationships with the third world that have structured life there in profound ways; they deny real social connections even as they evoke empathy."[17] *National Geographic*'s trademark portrayals of universal principles such as familial love, pride in labor and vigorous youth may allow for the reader's greater empathy, but such representations also provide constant reassurance of a stable, unthreatening world.

In its earlier decades, as I have discussed, the *National Geographic Magazine* spelled out the concept of cultural evolution in its written texts as well as its photographs and captions. In the second half of the twentieth century, *National Geographic* was still telling the tale of cultural evolution, if more subtly. Lutz and Collins detail several ways in which the photographs form a narrative of cultural evolution, equating more complicated technology with greater progress and civilization, and countering the ritual/traditional/indigenous conglomerate with that of science/modern/Western. Following *National Geographic*'s crude template of skin color as racial marker, they

14 Ibid.

15 Lutz and Collins, *Reading National Geographic*, 65.

16 Ibid.

17 Ibid., 280.

find that the darker a person's skin, the more likely that person is to be photographed in ritual practices and the less likely to be seen with machines.

National Geographic's world was an idealized world of entertaining difference, but it was also the world presented as ripe for capitalism. In the first fifteen years or so of America's newly discovered world, and *National Geographic*'s newly realized audience, the magazine varied in its representation of imperialism. Assessments of a place's resources – mineral, vegetable, and human – alternated with sentimental appeals to benevolent paternalism. By the 1920s, the magazine had settled on a more confident universalist hierarchy. This allowed Williams, in his "happy world" endeavors, to conflate his mission with that of the *Geographic* and allowed Adams to both sympathize with and champion indigenous peoples while heralding the ambitions of their conquerors.

Strategies of innocence require constant effort. Repetition and consistency help. The United States had its story of enlightened expansion told again and again in newspapers, in school textbooks, in juvenile literature, in magazines like the *National Geographic Magazine*. Overcoming isolationist reluctance to enter the two world wars helped burnish the savior image. Soon after Alexander Graham Bell hired Gilbert Grosvenor, he advised his young prospective son-in-law to "Hitch your wagon to a star."[18] Grosvenor did indeed, hitching his wagon to the rising star of the United States. As the United States established itself as a "global leader," *National Geographic* established itself as a "circulation leader."

National Geographic created and burnished its own exceptionalism as it promoted the exceptionalism of the United States. The world that *National Geographic* presented to its readers was indeed a world in which the United States stood as earnest leader in the "march of progress." It was a world of friendly faces and inviting places, no matter how remote or culturally or climatically exotic in comparison with the United States. *National Geographic* wanted as many readers as possible; the world it gave its American audience was a world that they wanted to see.

18 AGB to GHG, 12 July 1899, Box 99, Grosvenor Papers.

Bibliography

Manuscript sources

Alexander Graham Bell Family Papers, Manuscript Division, Library of Congress, Washington, D.C.

Grosvenor Family Papers, Manuscript Division, Library of Congress, Washington, D.C.

Records of the Society of Woman Geographers, Manuscript Division, Library of Congress, Washington, D.C.

National Geographic Society Archives, National Geographic Society, Washington, D.C.

Simms Family Papers, Manuscript Division, Moorland-Spingarn Research Center, Howard University, Washington, D.C.

Maynard Owen Williams Manuscript Collection, Kalamazoo College, Kalamazoo, Michigan.

Weimer K. Hicks Papers, Kalamazoo College, Kalamazoo, Michigan.

Harriet Chalmers Adams Scrapbooks, Stockton Public Library, Stockton, California. The scrapbooks contain many newspaper clippings that are referred to in footnote references but not in the bibliography below.

Transcript, Melville Bell Grosvenor Oral History Interview I, 28 April 1969, by Joe B. Frantz, LBJ Library. Online: ftp://webstorage2.mcpa.virginia.edu/library/nara/lbj/oralhistory/grosvenor_melville_1969_0428.pdf, accessed 28 July 2006.

Books, articles and additional scholarly writings

Abramson, Howard S. *National Geographic: Behind America's Lens on the World.* New York: Crown Publishers, 1987.

Adams, Harriet Chalmers. "Picturesque Paramaribo." *National Geographic Magazine* 43 (June 1907): 365–373.

______. "The East Indians in the New World (Trinidad)." *National Geographic Magazine* 43 (July 1907): 485–491.

______. "Some Wonderful Sights in the Andean Highlands," *National Geographic Magazine* 19 (September 1908): 597–618.

______. "Cuzco, America's Ancient Mecca." *National Geographic Magazine* 19 (October 1908): 669–689.

______. "The First Transandine Railroad from Buenos Aires to Valparaiso." *National*

Geographic Magazine 21 (May 1910): 397–417.

______. "Snapshots of Philippine America." *The World's Work* (May 1914): 31–42.

______. "Uncle Sam's White Magic." *World Outlook* (April 1915).

______ "In French Lorraine: That Part of France Where the First American Soldiers Have Fallen." *National Geographic Magazine* 32 (November–December 1917): 499–518.

______. "Rio de Janeiro, In the Land of the Lure," *National Geographic Magazine* 38 (September 1920): 165–210.

______. "A Longitudinal Journey Through Chile." *National Geographic Magazine* 42 (September 1922): 219–273.

______. "The Truth About Spain and Primo de Rivera." *The American Review of Reviews* 71 (January 1925): 69–72.

______. "Across French and Spanish Morocco." *National Geographic Magazine* 47 (March 1925): 327–356.

______. "Cirenaica, Eastern Wing of Italian Libia." *National Geographic Magazine* 57 (June 1930): 689–726.

______. "River-Encircled Paraguay." *National Geographic Magazine* 63 (April 1933): 385–416.

Adams, M.P. Greenwood. "Australia's Wild Wonderland." *National Geographic Magazine* 45 (March 1924): 329–356.

Alloula, Malek. *The Colonial Harem*. Minneapolis: University of Minnesota Press, 1986.

American Social History Project. *Savage Acts: Wars, Fairs and Empire*. Pennee Bender, Joshua Brown, and Andrea Ades Vasquez, directors. Center for Media and Learning, City of the University of New York, 1995.

Anderson, Benedict. *Imagined Communities: Reflections on the Origin and Spread of Nationalism*, 2nd ed. New York: Verso, 1991.

Anderson, Judith Icke. *William Howard Taft: An Intimate History*. New York: W. W. Norton, 1981.

Appleton, Louise. "Distillations of Something Larger: The Local Scale and American National Identity." *Cultural Geographies* 9 (2002): 421–447.

Atkinson, Edward. "Some Lessons in Geography." *National Geographic Magazine* 16 (April 1905): 193–198.

Austin, O. P. "Colonial Systems of the World." *National Geographic Magazine* 10 (January 1899): 21–25.

Babbitts, Judith. "Half-Tones and Half Truths: One Hundred Years of American Photographs of Japan." Paper presented to the American Studies Association Annual Meeting, Pittsburgh, November 1995.

Banta, Martha. *Imaging American Women: Idea and Ideals in Cultural History*. New York: Columbia University Press, 1987.

Banta, Melissa and Curtis M. Hinsley. *From Site to Sight: Anthropology, Photography, and the Power of Imagery*. Cambridge, MA: Peabody Museum Press, 1986.

Barnes, Albert C. *The Art in Painting*, revised ed. New York: Harcourt, Brace and Company, 1937.

Barthes, Roland. *Mythologies*. Translated by A. Lavers. New York: Hill and Wang, 1972.

______. *Camera Lucida: Reflections on Photography*. Translated by Richard Howard. New York: Hill & Wang, 1981.

Bartlett, Charles H. "Untoured Burma." *National Geographic Magazine* 24 (July 1913): 835–853.

Barton, Thomas F. and P. P. Karan. *Leaders in American Geography, vol. 1: Geographic Education*. Mesilla, New Mexico: New Mexico Geographical Society, 1992.

Bell, Alexander Graham. "Address of the President of the National Geographic Society to the Board of Managers." *National Geographic Magazine* 11 (October 1899): 401–408.

Bell, Morag and Cheryl McEwan, "The Admission of Women Fellows to the Royal Geographical Society, 1892–1914: the Controversy and the Outcome." *The Geographical Journal* 162 (November 1996): 295–312.

Bell, Morag, Robin A. Butlin and Michael Heffernan, eds. *Geography and Imperialism, 1820–1940*. Manchester: Manchester University Press, 1995.

Berger, John. *Ways of Seeing*. London and Harmondsworth, British Broadcasting Corporation and Penguin Books, 1972.

______. "Understanding a Photograph." In *Classic Essays in Photography*, edited by Alan Trachtenberg, 291–294. New Haven: Leete's Island Books, 1980.

Bingham, Alfred M. "Raiders of the Lost City." *American Heritage* 38 (July–August 1987): 54–64.

Bird, Jr., William L. "A Suggestion Concerning James Smithson's Concept of 'Increase and Diffusion.'" *Technology and Culture* 24 (April 1983): 246–255.

Bloom, Lisa. *Gender on Ice: American Ideologies of Polar Expeditions*. Minneapolis: University of Minnesota Press, 1993.

Blunt, Alison. *Travel, Gender and Imperialism: Mary Kingsley and West Africa*. New York: Guilford Press, 1994.

Brands, H. W. *Bound to Empire: The United States and the Philippines*. New York: Oxford University Press, 1992.

Bryan, C. D. B. *The National Geographic Society: 100 Years of Adventure and Discovery*. New York: Harry N. Abrams, 1987.

Buckley, Tom. "With The National Geographic on Its Endless, Cloudless Voyage." *New York Times Magazine*, 6 September 1970.

Bunkse, Edmunds. "Humboldt and an Aesthetic Tradition in Geography." *Geographical Review* 71 (1981): 127–46.

Burnett, D. Graham. *Masters of All They Surveyed: Exploration, Geography and a British El Dorado*. Chicago: University of Chicago Press, 2000.

Burns, Edward McNall. *The American Idea of Mission: Concepts of National Purpose and Destiny*. New Brunswick: Rutgers University Press, 1957.

Cagan, Steve. "Photography's Contribution to the 'Western' Vision of the Colonized 'Other.'" Paper presented at the Center for Historical Analysis, Rutgers University, 1990.

Castle, David Barton. "Leo Stanton Rowe and the Meaning of Pan Americanism." In *Beyond the Ideal: Pan Americanism in Inter-American Affairs*, edited by David Sheinin, 33–44. Westport, Connecticut: Praeger, 2000.

Chandler, Douglas "Changing Berlin," *National Geographic Magazine* 71 (February 1937): 131–177.

Chiari, Joseph. *Art and Knowledge*. New York: Gordian Press: 1977.

Church, John W. "A Vanishing People of the South Seas — The Tragic Fate of the Marquesan Cannibals, Noted For Their Warlike Courage and Physical Beauty." *National Geographic Magazine* 36 (1919): 275–306.

Clark, Kenneth. *The Nude: A Study in Ideal Form*. Garden City, New York: Doubleday, 1956.

Cohn, Jan. *Creating America: George Horace Lorimer and the* Saturday Evening Post. Pittsburgh: University of Pittsburgh Press, 1989.

Conrad, Joseph. "Geography and Some Explorers." *National Geographic Magazine* 45, (March 1924): 239–274.

Conroy, Sarah Booth. "She Painted the Town Pink." *Washington Post*, 1 February 1999, A1.

Cosgrove, Denis. *Social Formation and Symbolic Landscape*. London: Croom Helm, 1984.

Croce, Benedetto. *Guide to Aesthetics*. Translated by Patrick Romanell. New York: Bobbs-Merrill, 1965.

Crosby, Alfred W. *Ecological Imperialism: The Biological Expansion of Europe, 900–1900*. New York: Cambridge University Press, 1986.

Curtis, James. *Mind's Eye, Mind's Truth: FSA Photography Reconsidered.* Philadelphia: Temple University Press. 1989.

Davis, M. Kathryn. "The Forgotten Life of Harriet Chalmers Adams: Geographer, Explorer, Feminist." Masters thesis, Department of History, San Francisco State University, 1995.

Davis, William Morris. "Geography in the United States, I." *Science* NS 19 (January 1904): 121–132.

De Groot, Joanna. "'Sex' and 'Race': The Construction of Language and Image in the Nineteenth Century." In *Sexuality and Subordination*, edited by Susan Mendus and Jane Randall, 89–128. London: Routledge: 1989.

Dodge, Richard Elwood. *Dodge's Comparative Geography*, revised ed. Chicago: Rand McNally, 1912.

Domosh, Mona. "Toward a Feminist History of Geography." *Transactions of the Institute of British Geographers* N.S. 16 (1991): 95–104.

______. "Beyond the Frontiers of Geographical Knowledge." *Transactions of the Institute of British Geographers* N.S. 16 (1991): 488–490.

Dorfman, Ariel. *The Empire's Old Clothes: What the Lone Ranger, Babar, and Other Innocent Heroes Do to Our Minds*. New York: Penguin Books, 1983.

Drayton, Richard. *Nature's Government: Science, Imperial Britain, and the "Improvement" of the World*. New Haven: Yale University Press, 2000.

Driver, Felix. *Geography Militant: Cultures of Exploration and Empire.* Oxford: Blackwell, 2001.

______. "Geography's Empire: Histories of Geographic Knowledge." *Environment and Planning D: Society and Space* 10 (1992): 23–40.

Dubofsky, Melvin, Athan Theoharis and Daniel M. Smith. *The United States in the Twentieth Century*. Englewood Cliffs, NJ: Prentice-Hall, 1978.

Eber, Dorothy Harley. *Genius at Work*. New York: Viking, 1982.

Edney, Matthew H. *Mapping an Empire: The Geographic Construction of British India, 1765–1843*. Chicago: University of Chicago Press, 1997.

Edwards, Elizabeth. "Photographic 'Types': the Pursuit of Method." *Visual Anthropology* 3 (1990): 235–258.

______, ed. *Anthropology and Photography 1860-1920*. New Haven: Yale University Press, 1992.

Ehrick, Christine. "*Madrinas* and Missionaries: Uruguay and the Pan American Women's Movement," *Gender and History* 10(3), November 1998: 406–424.

Ekirch, Arthur A. Jr. *Progressivism in America: A Study of the Era from Theodore Roosevelt to Woodrow Wilson*. New York: New Viewpoints, 1974.

Emerson, Ralph Waldo. *English Traits* (excerpt), in *The Norton Book of Travel*, edited by Paul Fussell, 362–380. New York: W.W. Norton & Co., 1987.

Fairchild, David. "The Jungles of Panama." *National Geographic Magazine* 41 (February 1922): 131–145.

______. *The World Was My Garden*. (New York: Scribner's Sons, 1938)

Farhi, Paul. "Mapping Out a Greater Society; National Geographic Is Pushing Beyond Its Yellow Borders Into New Market Territory." *The Washington Post*, 27 January 1997, F12.

Flack, James K. *Desideratum in Washington: The Intellectual Community in the Capital City, 1870-1900*. Cambridge, Massachusetts: Schenkman, 1975.

Fogelsong, Richard E. *Planning the Capitalist City: The Colonial Era to the 1920s*. Princeton: Princeton University Press, 1986.

Forbes, Edgar Allen. "Notes on the Only American Colony in the World." *National Geographic Magazine* 21 (1910): 719–29.

Freeman, T. W. *A Hundred Years of Geography*. London: Gerald Duckworth & Co., 1961.

Fry, Joseph A. "Imperialism, American Style, 1890-1916." In *American Foreign Relations Reconsidered, 1890-1993*, edited by Gordon Martel, 52–70. New York: Routledge, 1994.

Gannett, Henry. "The Annexation Fever." *National Geographic Magazine* 8 (December 1897): 354-358.

Garner, J. W. "The Carrie Chapman Catt Citizen Course: The United States and Its Dependencies: Uncle Sam's Step-Children." *The Woman Citizen* (August 21, 1920): 312–313, 322.

______. "The Carrie Chapman Catt Citizen Course: Some More About Our Dependencies; How We Came By Them and How They Are Governed." *The Woman Citizen* (August 28, 1920): 340–342.

Gascoigne, John. *Science in the Service of Empire: Joseph Banks, the British State and the Uses of Science in the Age of Revolution*. New York: Cambridge University Press, 1998.

Gilman, Sander L. "Black Bodies, White Bodies: Toward an Iconography of Female Sexuality in Late Nineteenth Century Art, Medicine, and Literature." *Critical Inquiry* 12 (Autumn 1985): 204–42.

Glacken, Clarence J. *Traces on the Rhodian Shore: Nature and Culture in Western Thought from Ancient Times to the End of the Eighteenth Century*. Berkeley: University of California Press, 1967.

Glueck, Grace. "With an Eye for Nature and Its Exquisite Forms." *The New York Times*, 16 August 1996, C30.

Godlewska, Anne and Neil Smith, eds. *Geography and Empire*. Oxford: Blackwell, 1994.

Gould, Stephen Jay. *The Mismeasure of Man*. New York: W.W. Norton and Company, 1981.

Graham-Brown, Sarah. *Image of Women: The Portrayal of Women in Photography of the Middle East 1860–1950*. New York: Columbia University Press, 1988.

Gramsci, Antonio. *Selections from the Prison Notebooks*. London: Lawrence & Wishart, 1971.

Gray, Allison. "We All Have a Secret Love of Adventure and Romance." *The American Magazine* (May 1922): 26–29, 133–135.

Green, David. "Classified Subjects: Photography and Anthropology: The Technology of Power," *Ten-8* 14 (1984): 30–37.

Greenblatt, Stephen. *Marvelous Possessions: The Wonder of the New World*. Chicago: University of Chicago Press, 1991.

Gregory, Derek. *Geographical Imaginations*. Cambridge, MA: Blackwell, 1994.

Grosvenor, Gilbert H. "A Revelation of the Filipinos." *National Geographic Magazine* 16 (April 1905): 139–92.

______. "Young Russia: The Land of Unlimited Possibilities." *National Geographic Magazine* 26 (November 1914): 421–520.

______. "Report of the Director and Editor of the National Geographic Society for the Year 1914." *National Geographic Magazine* 27 (March 1915): 318–320.

______. "The Hawaiian Islands — America's Strongest Outpost of Defense — The Volcanic and Floral Wonderland of the World." *National Geographic Magazine* 45 (February 1924): 115–238.

______. *The National Geographic Society and Its Magazine*. Washington, D.C.: National Geographic Society, 1957.

______. "The Story of the Geographic." In *National Geographic Index 1888–1946*. Washington, D.C.: National Geographic Society, 1967.

Guy, Donna J. "The Politics of Pan American Cooperation: Maternalist Feminism and the Child Rights Movement, 1913–1960." *Gender & History* 10 (November 1998): 449–469.

"Haiti and Its Regeneration by the United States." *National Geographic Magazine* 38 (December 1920): 497–511.

Hall, Stuart. "Gramsci's Relevance for the Study of Race and Ethnicity." In *Stuart Hall: Critical Dialogues in Cultural Studies*, edited by David Morley and Kuan-Hsing Chen, 411–440. London: Routledge, 1996.

Haller, Mark H. *Eugenics: Hereditarian Attitudes in American Thought.* New Brunswick: Rutgers University Press, 1984.

Halpin, Zuleyma Tang. "Scientific Objectivity and the Concept of the 'Other.'" *Women's Studies International Forum*. 12(3) (1989): 285–94.

Hannaford, Ivan. *Race: The History of an Idea in the West*. Washington, D.C.: Woodrow Wilson Center Press, 1996.

Hanson, Jim. *The Decline of the American Empire*. Westport, Connecticut: Praeger, 1993.

Haraway, Donna. *Primate Visions: Gender, Race and Nature in the World of Modern Science*. New York: Routledge, 1989.

Hays, Constance L. "Seeing Green in a Yellow Border: Quest for Profits is Shaking a Quiet Realm." *The New York Times*, 3 August 1997, sec. III, 1, 12–13.

Herman, Edward S. and Noam Chomsky. *Manufacturing Consent: The Political Economy of the Mass Media.* New York: Pantheon, 1988.

Hildebrand, J. R. "Machines Come to Mississippi." *National Geographic Magazine* 72 (September 1937): 263–318.

Hinsley, Curtis M. Jr. *Savages and Scientists: The Smithsonian Institution and the Development of American Anthropology 1846–1910*. Washington, D.C.: Smithsonian Institution Press, 1981.

Hofstadter, Richard. *Social Darwinism in American Thought*, revised ed. Boston: Beacon Press, 1955.

Hunt, Lynn A. *Politics, Culture and Class in the French Revolution.* Berkeley: University of California Press, 1984.

Hunt, Michael H. *Ideology and United States Foreign Policy*. New Haven: Yale University Press, 1987.

"Imagination and Geography," *National Geographic Magazine* 18 (December 1907), 825.

James, Preston E. *All Possible Worlds: A History of Geographical Ideas*. New York: Bobbs-Merrill, 1972.

Johnson, Martin. *Art and Scientific Thought: Historical Studies towards a Modern Revision of Their Antagonism*. New York: AMS Press, 1949.

Kaufman, Burton I. *Efficiency and Expansion: Foreign Trade Organization in the Wilson Administration, 1913–1921*. Westport, Connecticut: Greenwood Press, 1974.

Keen, Benjamin and Keith Haynes. *A History of Latin America* (Seventh Edition), Boston: Houghton Mifflin, 2004.

Kenney, Nathaniel T. "The Winds of Freedom Stir a Continent," *National Geographic Magazine* 118 (September 1960): 303–359.

Kolodny, Annette. *The Lay of the Land: Metaphor as Experience and History in American Life and Letters*. Chapel Hill: University of North Carolina Press, 1975.

Lacey, Michael James. "The Mysteries of Earth-Making Dissolve: A Study of Washington's Intellectual Community and the Origins of American Environmentalism in the Late Nineteenth Century." Ph.D. dissertation, George Washington University, 1979.

La Gorce, John Oliver. *The Story of the Geographic*. Washington, D.C.: James Wm. Bryan Press, 1915.

Landes, Joan B. *Women and the Public Sphere in the Age of the French Revolution*. Ithaca: Cornell University Press, 1988.

Lane, Franklin K. "A Mind's Eye Map of America," *National Geographic Magazine* 36 (June 1920): 479–518.

Lippmann, Walter. *Public Opinion*. New York: Free Press, 1965.

Livingstone, David N. *The Geographical Tradition: Episodes in the History of a Contested Enterprise*. Oxford: Blackwell, 1992.

Lutz, Catherine A. and Jane L. Collins. *Reading National Geographic*. Chicago: University of Chicago Press, 1993.

Macdermott, Isabel Keith. A Significant Pan American Conference. Washington, D.C.: Government Printing Office, 1922. Digital book, Harvard University Library Page Delivery Service http://ocp.hul.harvard.edu/ww/outsidelink.html/ http://nrs.harvard.edu/urn-3:FHCL:479113, accessed 30 July 2006).

Mann, William M. *Wild Animals In and Out of the Zoo*. Washington, D.C.: Smithsonian Institution, 1930.

McCarry, Charles. "Three Men Who Made the Magazine." *National Geographic* 174 (September 1988): 287–316.

McClintock, Anne. *Imperial Leather: Race, Gender and Sexuality in the Colonial Context*. New York: Routledge, 1995.

McGee, W J. "The Work of the National Geographic Society," *National Geographic Magazine* 7 (August 1896), 253–259.

______. "Fifty Years of American Science," *The Atlantic Monthly* (May 1898).

______. "American Geographic Education." *National Geographic Magazine* 9 (July 1898): 305–307.

______. "The Growth of the United States." *National Geographic Magazine* 9 (September 1898): 377–386.

______. "National Growth and National Character." *National Geographic Magazine* 10 (June 1899): 185–206.

Merchant, Carolyn. *The Death of Nature: Women, Ecology and the Scientific Revolution*. New York: Harper & Row, 1980.

Miller, Mark J. "To the Ends Of The Earth." *Folio*, 1 March 2004, 3.

Mills, Sara. *Discources of Difference: An Analysis of Women's Travel Writing and Colonization*. London: Routledge, 1991.

Mitchell, Timothy. *Colonizing Egypt*. Cambridge: Cambridge University Press, 1988.

Mitchell, W. J. T. *Picture Theory: Essays on Verbal and Visual Representation.* Chicago: University of Chicago Press, 1994.

Moore, W. Robert. "Among the Hill Tribes of Sumatra." *National Geographic Magazine* 57 (1930): 187–227.

Moritzen, Julius. *The Peace Movement of America.* New York: G. P. Putnam Sons. 1912.

Mott, Frank Luther. *American Journalism: A History: 1690–1960*, 3rd ed. New York: Macmillan, 1962.

"National Geographic Gets A Facelift." *The Toronto Star*, 21 August 2000.

Nelson, Edward William. "Henry Wetherbee Henshaw, Naturalist 1850–1930," *The Auk: A Quarterly Journal of Ornithology* 49 (October 1932): 399–427.

Normano, J. F. *The Struggle for South America: Economy and Ideology.* Boston: Houghton Mifflin Company, 1931.

"Nicaragua, Largest of Central American Republics." *National Geographic Magazine* 51 (March 1927), 370–378.

Ohmann, Richard. *Selling Culture: Magazines, Markets and Class at the Turn of the Century.* London and New York: Verso, 1996.

Osgood, Robert Endicott. *Ideals and Self-Interest in America's Foreign Relations: The Great Transformation of the Twentieth Century.* Chicago: University of Chicago Press, 1953.

"Our Imperialist Propaganda: The *National Geographic*'s Anti-Haitian Campaign," *The Nation* 112 (6 April 1921). In *Anti-Imperialism in the United States, 1898-1935*, edited by Jim Zwick, http://www.boondocksnet.com/ai/ailtexts/nation210406.html (accessed 26 February 2004).

Pateman, Carole. *The Sexual Contract.* Stanford: Stanford University Press, 1988.

Patterson, Carolyn Bennett. *Of Lands, Legends, and Laughter: The Search for Adventure with National Geographic.* Golden, Colorado: Fulcrum Publishing, 1998.

Pauly, Philip. "The National Geographic Society and the Iconography of an Emerging World." Paper, Johns Hopkins University, 1976.

______. "The World and All That Is In It: The National Geographic Society, 1888–1918." *American Quarterly* 31 (Fall 1979): 517–32.

Peterson, Theodore. *Magazines in the Twentieth Century*, 2nd ed. Urbana: University of Illinois Press, 1964.

Phillips, Richard. *Mapping Men and Empire: A Geography of Adventure.* London: Routledge, 1997.

Pinney, Christopher. "Classification and Fantasy in the Photographic Construction of Caste and Tribe." *Visual Anthropology* 3 (1990): 259–287.

Poignant, Roslyn. "Surveying the Field of View: The Making of the RAI Photographic Collection." In *Anthropology and Photography 1860–1920*, edited by Elizabeth Edwards, 43–73. New Haven: Yale University Press, 1992.

Poole, Robert M. *Explorers House: National Geographic and the World It Made*, New York: Penguin, 2004.

Pratt, Julius W. *The Expansionists of 1898: The Acquisition of Hawaii and the Spanish Islands*. Baltimore: Johns Hopkins Press, 1936; reprint ed., Gloucester, MA: Peter Smith, 1959.

Pratt, Mary Louise. *Imperial Eyes: Travel Writing and Transculturation*. New York: Routledge, 1992.

Prisco, Salvatore. "John Barrett and Collective Approaches to United States Foreign Policy in Latin America, 1907–20." *Diplomacy & Statecraft* 14 (September 2003), 57–69.

Raeburn, John. "African-Americans Mediated: The Photo League, the 'Harlem Document,' *Native Son*, and *Look* Magazine." Paper presented at the American Studies Association annual meeting, Kansas City, 1996.

Rafael, Vicente L. "White Love: Surveillance and Nationalist Resistance in the U.S. Colonization of the Philippines." In *Cultures of United States Imperialism*, edited by Amy Kaplan and Donald E. Pease, 185–218. Durham, NC: Duke University Press, 1993.

Riggs, Marlon. *Ethnic Notions*. San Francisco: California Newsreel, 1986.

Roller, Arnold. "Pan-American Union?" *The Nation* 126 (18 January 1928): 78–80

Rose, Gillian. *Visual Methodologies* (London: Sage Publications, 2001).

Rosenbaum, Ron. "The Great Ivy League Nude Posture Photo Scandal." *The New York Times Magazine*, 15 January 1995: 26–31, 40, 46, 55–56.

Rosenberg, Emily S. *Spreading the American Dream: American Economic and Cultural Expansion 1890–1945*. New York: Hill & Wang, 1982.

Ross, Ishbel. "Geography, Inc.," *Scribner's Magazine* 103 (June 1938): 23–27, 57.

Rothenberg, Tamar. "Voyeurs of Imperialism: *The National Geographic Magazine* Before World War II." In *Geography and Empire*, edited by Anne Godlewska and Neil Smith, 155–172. Oxford: Basil Blackwell, 1994.

Ryan, James R. *Picturing Empire: Photography and the Visualization of the British Empire*. Chicago: University of Chicago Press, 1997.

Ryan, Mary P. *Women in Public: Between Banners and Ballots, 1825–1880*. Baltimore: Johns Hopkins University Press, 1990.

Scheibinger, Londa. *Nature's Body: Gender in the Making of Modern Science*. Boston: Beacon Press, 1993.

Schirmer, Daniel B. *Republic or Empire: American Resistance to the Philippine War*. Cambridge, Massachusetts: Schenkman, 1972.

Schneirov, Matthew. *The Dream of a New Social Order: Popular Magazines in America 1893–1914*. New York: Columbia University Press, 1994.

Schulten, Susan. *The Geographical Imagination in America, 1880–1950*. Chicago: University of Chicago Press, 2001.

Sekula, Allan. "The Body and the Archive." In *The Contest of Meaning: Critical Histories of Photography*, edited by Richard Bolton, 343–388. Cambridge: MIT Press, 1989.

Sharp, Joanne P. *Condensing the Cold War:* Reader's Digest *and American Identity*. Minneapolis: University of Minnesota Press, 2000.

Shay, Felix. "Cairo to Cape Town, Overland." *National Geographic Magazine* 47 (February 1925):123–260.

Sheinin, David, ed., *Beyond the Ideal: Pan Americanism in Inter-American Affairs*. Westport, Connecticut: Praeger, 2000.

Slotkin, Richard. *Lost Battalions: The Great War and the Crisis of American Nationality*. New York: Henry Holt, 2005.

Smith, Bernard. *European Vision and the South Pacific*. Oxford University Press, 1960.

Smith, Henry Nash. *Virgin Land: The American West as Symbol and Myth*. Cambridge: Harvard University Press, 1950.

Smith, Joseph. "The First Conference of American States (1889-1890) and the Early Pan American Policy of the United States." In *Beyond the Ideal: Pan Americanism in Inter-American Affairs*, edited by David Sheinin, 19–32. Westport, Connecticut: Praeger, 2000.

Sontag, Susan. *On Photography*. New York: Farrar, Straus and Giroux, 1973.

Spencer, Frank. "Some Notes on the Attempt to Apply Photography to Anthropometry During the Second Half of the Nineteenth Century." In *Anthropology and Photography 1860–1920*, edited by Elizabeth Edwards, 99–107. New Haven: Yale University Press, 1992.

Stafford, Barbara Maria. *Voyage into Substance: Art, Science, Nature, and the Illustrated Travel Account, 1760–1840*. Cambridge, MA: The MIT Press, 1984.

Stafford, Robert A. *Scientist of Empire: Sir Roderick Murchison, Scientific Exploration and Victorian Imperialism*, Cambridge: Cambridge University Press, 1989.

Steet, Linda. *Veils and Daggers: A Century of* National Geographic*'s Representation of the Arab World*. Philadelphia: Temple University Press, 2000.

Stepan, Nancy. *The Idea of Race in Science: Great Britain 1800–1960*. London: Macmillan, 1982.

______. "Race and Gender: The Role of Analogy in Science," *Isis* 77 (June 1985): 261–277.

Stewart, Susan. "Death and Life, In That Order, In the Works of Charles Willson Peale." In *Visual Display: Culture Beyond Appearances*, edited by Lynne Cooke and Peter Wollen, 31–53. Seattle: Bay Press, 1995.

Stocking, Jr., George W. *Race, Culture and Evolution: Essays in the History of Anthropology*. New York: Free Press, 1968.

______. "The Camera Eye as I Witness: Skeptical Reflections on the 'Hidden Messages' of *Anthropology and Photography, 1860-1920*." *Visual Anthropology* 6 (1993): 211–218.

Stoddart, D. R. *On Geography and Its History*. New York: Basil Blackwell, 1986.

Stoler, Ann. "Making Empire Respectable: The Politics of Race and Sexual Morality in 20th Century Colonial Cultures," *American Ethnologist* 16 (November 1989): 634–60.

Stoner, K. Lynn. "In Four Languages But with One Voice: Division and Solidarity within Pan American Feminism, 1923–1933." In *Beyond the Ideal: Pan*

Americanism in Inter-American Affairs, edited by David Sheinin, 79–94. Westport, Connecticut: Praeger, 2000.

Taft, William H. "Ten Years in the Philippines." *National Geographic Magazine* 19 (February 1908): 141–148.

Tarbell, Ida M. *All in a Day's Work: An Autobiography*. Urbana and Chicago: University of Illinois Press, 2003.

Thaw, Lawrence Copley and Margaret S. Thaw. "Trans-Africa Safari: A Motor Caravan Rolls Across Sahara and Jungle Through Realms of Dusky Potentates and the Land of Big-Lipped Women," *National Geographic Magazine* 74 (September 1938): 327–364.

Thomas, Alan. *Time in a Frame: Photography and the Nineteenth-Century Mind*. New York: Schocken Books, 1977.

Thomas, Nicholas. *Colonialism's Culture: Anthropology, Travel and Government*. Princeton: Princeton University Press, 1994.

Tiffany, Sharon W. and Kathleen J. Adams. *The Wild Women: An Inquiry into the Anthropology of an Idea*. Cambridge, Massachusetts: Schenkman, 1985.

Torgovnick, Marianna. *Gone Primitive: Savage Intellects, Modern Lives*. Chicago: University of Chicago Press, 1990.

Trachtenberg, Alan. *Reading American Photographs: Images as History, Mathew Brady to Walker Evans*. New York: Hill and Wang, 1989.

Tuason, Julie A. "The Ideology of Empire in *National Geographic Magazine*'s Coverage of the Philippines, 1889–1908." *The Geographical Review* 89 (1) (January 1999): 34–53.

Veeser, Cyrus. *A World Safe for Capitalism: Dollar Diplomacy and America's Rise to Global Power*. New York: Columbia University Press, 2002.

Vesilind, Priit Juho. "The Development of Color Photography at *National Geographic*." M.A. thesis, Syracuse University, 1977.

Wach, Howard M. "'Expansive Intellect and Moral Agency': Public Culture in Antebellum Boston." *Proceedings of the Massachusetts Historical Society* 107 (1995): 30–56.

______. "Culture and the Middle Classes: Popular Knowledge in Industrial Manchester," *Journal of British Studies* 27 (October 1988): 375–404.

Wallace, Frederick William. "Life on the Grand Banks: An Account of the Sailor-Fishermen Who Harvest the Shoal Waters of North America's Eastern Coasts," *National Geographic Magazine* 40 (July 1921): 1–28.

Ward, Robert De Courcy. "Our Immigration Laws from the Viewpoint of Natural Eugenics." *National Geographic Magazine* 23 (January 1912): 38–41.

Ware, Vron. *Beyond the Pale: White Women, Racism and History*. London: Verso, 1992.

Warner, Marina. *Monuments and Maidens: The Allegory of the Female Form*. New York: Atheneum, 1985.

Watkins, T. H. "The Greening of the Empire: Sir Joseph Banks." *National Geographic* 190 (November 1996): 28–52.

Weber, Max. *The Methodology of the Social Sciences*. Translated and edited by Edward A. Shils and Henry A. Finch. Glencoe, Illinois: The Free Press, 1949.

Wharton, Elna Harwood. "A Woman Turns Geographer," *The Forecast*, July 1930.

Whitaker, Arthur P. *The Western Hemisphere Idea: Its Rise and Decline*. (Ithaca: Cornell University Press, 1954).

Wiarda, Howard J. *Cracks in the Consensus: Debating the Democracy Agenda in U.S. Foreign Policy*. Westport, CT: Praeger, 1997.

Wiebe, Robert H. *The Search for Order: 1877–1920*. New York: Hill and Wang, 1967.

Wilbur, Lyman D. "Surveying Through Khoresm: A Journey into Parts of Asiatic Russia Which Have Been Closed to Western Travelers Since the World War." *National Geographic Magazine* 61 (June 1932): 753–780.

Wilkerson, Marcus M. *Public Opinion and the Spanish-American War: A Study in War Propaganda*. New York: Russell & Russell, 1967 (1932).

Williams, Maynard Owen. "Russia's Orphan Races: Picturesque Peoples Who Cluster on the Southeastern Borderland of the Vast Slav Dominions." *National Geographic Magazine* 34 (October 1918): 245–278.

______. "Between Massacres in Van." *National Geographic Magazine* 36 (August 1919): 181–184.

______. "Adventures with a Camera in Many Lands." *National Geographic Magazine* 40 (July 1921): 87–112.

______. "At the Tomb of Tutankhamen: An Account of the Opening of the Royal Egyptian Sepulcher Which Contained the Most Remarkable Funeral Treasures Unearthed in Historic Times." *National Geographic Magazine* 43 (May 1923): 461–508.

______. "Latvia, Home of the Letts." *National Geographic Magazine* 46 (October 1924): 401–443.

______. "The First Natural-Color Photographs from the Arctic." *National Geographic Magazine* 49 (March 1926): 301–316.

______. "In the Birthplace of Christianity." *National Geographic Magazine* 50 (December 1926): 697–720.

______. "Color Records From the Changing Life of the Holy City." *National Geographic Magazine* 52 (December 1927): 682–707.

______ ."Unspoiled Cyprus: The Traditional Island Birthplace of Venus Is One of the Least Sophisticated of Mediterranean Lands." *National Geographic Magazine* 54 (July 1928): 1–55.

______. "New Greece, the Centenarian, Forges Ahead." *National Geographic Magazine* 58 (December 1930): 649–721.

______. "The Citroën Trans-Asiatic Expedition Reaches Kashmir." *National Geographic Magazine* 60 (October 1931): 387–443.

______. "First Over the Roof by Motor: The Trans-Asiatic Expedition Sets New Records for Wheeled Transport in Scaling Passes of the Himalayas." National Geographic Magazine 61 (March 1932): 321–363.

______. "From the Mediterranean to the Yellow Sea by Motor." *National Geographic*

Magazine 62 (November 1932): 513–580.

______. "Netherlands Indies: Patchwork of Peoples." *National Geographic Magazine* 73 (June 1938): 681–712.

______. "Bali and Points East: Crowded, Happy Isles of the Flores Sea Blend Rice Terraces, Dance Festivals and Amazing Music in Their Pattern of Living." *National Geographic Magazine* 75 (March 1939): 313–352.

______. "Mother Volga Defends Her Own." *National Geographic Magazine* 82 (December 1942): 793–811.

Williams, Raymond. "Selections from *Marxism and Literature*." In *Culture/Power/History*, edited by Nicholas B. Dirks, Geoff Eley and Sherry Ortner, 585–608. Princeton: Princeton University Press, 1994.

Wright, John Kirtland. *Geography in the Making: The American Geographical Society 1851–1951*. New York: American Geographical Society, 1952.

Young, Robert J.C. *Colonial Desire: Hybridity in Theory, Culture and Race*. London and New York: Routledge, 1995.

Website sources

www.bbc.co.uk/history/scottishhistory/europe/oddities_europe.shtml, accessed 16 July 2006.

http://en.wikipedia.org/wiki/Graham_Fairchild, accessed 22 August 2006.

www.explorenorth.com/library/bios/scidmore.html, accessed 21 April 2006.

www.ladnerslanding.com/Newsletters/Vol_2/March_2002.html, accessed April 21, 2006.

www.nationalgeographic.com/egypt, accessed October 21, 2006.

Index

National Geographic Magazine is abbreviated to *NGM* in the index, except for its own main entry)

References to illustrations are in bold.

Abramson, Howard 58
  NGM 19
Adams, Franklin Pierce 133, 134, 146, 162
  work with PAU 154
Adams, Harriet Chalmers 17, 18, 21, 89, 131-64
  on America-South America trade 155-7
  on cultural loss 161
  early life 132-3
  femininity
    representations 146-7
    use 144, 146
  feminism 148-50, 163
  on history 158, 160
  humor, in writings 142-3
  lectures 135, 144, 146
  National Geographic Society, relationship 137-8
  photographs by **149**, **152**, **159**
  photographs of **136**, **145**
  photography 134
  on race 161-2
  as 'seeing-man' 163
  social circle 135, 137
  travels 18, 151-3
    South America 131, 133-4, 152-3
  war correspondent 134, 148
  women explorers, views on 140-1
Adams, Kathleen J. 14
Addams, Jane 150
African-Americans, Maynard Williams on 117-18
Agassiz, Louis 87
Aguinaldo, Emilio 55
Aikman, Lonnelle 139
Alloula, Malek 92
America
  expansionism, and *NGM* 22, 169
  as imagined community 4-5
  scientific societies 25-6
  South America, trade 155-7
  Western Hemisphere, interventions 155
  women's suffrage 148
American Geographical Society (AGS) 27
American identity
  formation 41-2
  and *NGM* 5, 66-7, 165-6
  and *Reader's Digest* 5
The American Magazine 13
American Museum of Natural History 20, 52, 74, 79
The American Review of Reviews 158
American Society for the Advancement of Science 30
Anderson, Benedict, *Imagined Communities* 41
anti-conquest narratives 7, 73-4
  Humboldt 73
  in *NGM* 9, 71
  science 7-8
  see also 'seeing man'
apartheid, Gilbert Grosvenor on 65
Appleton, Louise 5
Association of American Geographers 11, 66
  foundation 38-9
Atlantic Monthly 34

Banks, Joseph 69-70, 75
bare-breasted women
  and Maynard Williams 114-19
  in *NGM* 15-16, 22, 54, 69, **91**, 94, 116
Barrett, John 154, 155
Barrow, John, *Travels* 79
Barthes, Roland 106, 108
  Camera Lucida 81-2
Bell, Alexander Graham 11, 32
  eugenics, interest in 49

National Geographic Society, president 33-6, 39
Berger, John 102
Bertillon, Alphonse 85-6
Bloom, Lisa, *Gender on Ice* 20
Blunt, Alison 142
Boston Herald, on the *NGM* 42-3
Boston Society for the Diffusion of Useful Knowledge 27
Bowdoin ship **101**
Bryan, C.D.B 19
Buffon, Comte de, *Natural History* 77
Burton, Richard 6, 100-1

Cagan, Steve 95
Catt, Carrie Chapman 150
Century 11, 34, 35, 78
Chandler, Douglas 58-9
Chapman, Frank M. 138
Christian Herald 99, 103, 122
Citroën-Haardt Trans-Asia Expedition 99, 105, 106, 125
Clark, Kenneth 78
Collins, Jane, *Reading National Geographic* (co-author) 21-2, 96, 106, 168
Conrad, Joseph, 'Geography and Some Explorers' 9
consent, manufacturing of 3-4
conservation
  and eugenics 50
  National Geographic Society 50
conservation movement 50-1
Cook, Capt James 9, 71
Cook, Frederick 19
Coolidge, Calvin 46
Cosmopolitan 4, 42
criminology, and photography 85-6
Cuba 155
  in *NGM* 32

Darwinism, Richard Hofstadter on 47
Davis, Kate 138, 147, 150
Davis, William Morris 28, 38
de la Blanche, Vidal 73
Defoe, Daniel, *Robinson Crusoe* 52
Densmore, Frances 145
Didion, Joan 7
Dominican Republic 155
Domosh, Mona 141
Dorfman, Ariel 10
Driver, Felix 24, 26
  Geography Militant 8
Dutch East Indies 111

Earhart, Amelia 139
Ellis, Havelock 85
Emerson, Ralph Waldo, *English Traits* 80
English-Speaking Union 48
eugenics
  Alexander Graham Bell's interest in 49
  and conservation 50
  and immigration 49
  in *NGM* 49
exploration
  gender bias 140-4, 146-53, 156
  and masculinity 141-2
  narratives of 74-5
  vs traveling 143-4
explorers, women, Mary Pratt on 142
Explorer's Club 18, 151

Fairchild, David 52
feminism, Harriet Adams 148-50, 163
Fisher, Franklin L. 17, 94, 95, 110
Flores, Pastoriza 150
Forster, Johann Reinhold 77
Foucault, Michel 86
Fry, Joseph A. 9

Galton, Francis 85
Gannett, Henry 32, 67
Garner, J.W. 53
gender bias
  in exploration 140-4, 146-53, 156
  National Geographic Society 138-40
geographical societies, establishment 27
geography
  as cultural product 23
  and imperialism 8-9
  presentation 24
  representation of 2
  as science 28, 71-2, 166
Gilder, Richard Watson 34
Graham, Maria 142
Gramsci, Antonio 3
Grant, Madison 49-50
Great Blizzard (1888) 28
Green, David 81, 86

Greenblatt, Steven 12
Gregory, Derek 71-3
Grosvenor, Edwin A. 33
Grosvenor, Edwin P. 33
Grosvenor, Elsie Bell 135
Grosvenor, Gilbert H. 7, 13, 16-17, 19, 32, 36-7, 41, 90, 103-4, 119
  anti-black bias 57
  anti-communism 59-60
  anti-semitism 59
  on apartheid 65
  editor, *NGM* 12, 35, 38, 39, 45-6, 47-8, 50, 54-5, 166
  fascism 59
  investment acumen 60
  politics 57-67, 104
Grosvenor, Melville Bell 65, 126
Gulf War (1991) 3-4

Haile Selassie 147
Haiti 155
  in *NGM* 63-4
Hall, Stuart 3
Halpin, Zuleyma Tang 74
Haraway, Donna 50, 74, 79
  'Teddy Bear Patriarchy' 20, 46, 52-3
Harper's Monthly 42
Harrison, Marguerite 151
Hawaii, annexation 29
hegemony
  concept 3
  as process 16
Henshaw, Henry W. 93-4
history, Harriet Adams on 158, 160
Hofstadter, Richard, on Darwinism 47
Holmes, Burton 116
Hubbard, Gardiner Greene 11, 32-3
Humboldt, Alexander von 6
  anti-conquest narratives 73
Huxley, Thomas Henry 85
Hyde, John 28, 32, 37-8

ideal
  nude as 78
  search for 78
identity, and otherness 5, 111
  see also American identity
imagined community
  America 4-5
  National Geographic Society 35
imperialism
  and geography 8-9
  and photography 13, 56, 71, 84
  and Progressivism 44-5
  travel writing as 6-7
Inter-American Commission of Women 150
Italian Riviera **102**
It's a Wonderful Life, *NGM* in 1-2

Jackson, W.H. 86
Japan, representations of 94
Johnson, Martin 14, 112
Johnson, President Lyndon B. 165

Kingsley, Mary 142, 147
Kolodny, Rochelle 82

La Gorce, John Oliver 17, 39, 57, 59, 63-4, 65, 90, 119, 126, 165
  on Harriet Adams 131-2, 146
  The Story of the Geographic 71
Ladies' Home Journal 4, 42, 148
Lamprey, John H. 85
League of Women Voters 150
liberal-developmentalism
  in *NGM* 157-8, 163
  Rosenberg on 45, 46-7
  tenets 45
  see also Progressivism
Linnaean taxonomy 76
Linnaeus, Carolus 75
  human beings, classification 76
Lippmann, Walter 3
Lombroso, Cesare 85
Lutz, Berta 150
Lutz, Catherine, *Reading National Geographic* (co-author) 21-2, 96, 106, 168

McClintock, Anne 12-13, 14
McClure, S.S. 38
McClure's 4, 34, 35, 42, 139
McGee, W.J. 25, 26, 30-1, 32
McKinley, President William 30
  assassination 44
magazines
  ideologies 4, 5
  imagined community 4-5
  mass culture, creation of 42

Maine battleship, destruction 29
Mann, William 112
maps, in *NGM* 93
Marquesa Islands 69, 70
masculinity
 and exploration 141-2
 ideal, *National Geographic Magazine* 52-3
Merrill, Fullerton 36
Metropolitan Museum of Art 70
Mexico 155
Mills, Sara 142, 143
minstrel shows 118
Monroe Doctrine, Corollary to (1904) 154-5
Moore, W. Robert 58
Munsey's 4, 35, 42, 78

narratives
 anti-conquest 9, 71, 73-4
 of exploration 74-5
The Nation 63, 64
National Geographic Magazine (*NGS*)
 advertisements, positioning 42
 and American expansionism 22, 169
 and American identity 5, 66-7, 165-6
 anti-communism 60-1
 anti-conquest narratives 9, 71
 anti-Russian bias 61-2
 bare-breasted women in 15-16, 22, 54, 69, **91**, 94, 116
 circulation 1, 65
 Cuba, issue on 32
 cultural influence 165
 as cultural marker 21
 eugenics in 49
 Flashback feature 69, 95, 167
 foreign editions 166
 guiding principles 62-3, 104
 Haiti in 63-4
 histories of 20-1
 in *It's a Wonderful Life* 1-2
 liberal-developmentalism in 157-8, 163
 maps in 93
 masculinity ideal 52-3
 membership, foreign 128
 multi-media formats 166, 167
 nude in 78
 Orientalism in 21, 166
 origins 1
 Panama in 51-2
 Philippines in 32, 54-6, 157-8
 photographers, instructions to 125-6
 and photography 13, 14, 22, 54-5, 69, 89-96, 168-9
  color 120-1
 political vision 43-54
 popularization 38, 39, 42
 present-day 166
 for professionals, phase 28
 readers' letters 23
 Republican bias 57
 Roosevelt contributions 44, 46
 Samoa in 70
 spin-offs 166, 167
 'strategies of innocence' 6, 7, 24, 169
 Taft contributions 46, 48-9
 typification in 80, 82, 89-96
 World Wars in 64-5
National Geographic Society 1, 2, 11, 12
 Alexander Graham Bell, president 33-4
 conservation 50
 exploration, funding of 137-8
 foundation 25
 gender bias 138-40
 histories of 19-20
 Hubbard medal 46
 imagined community 35
 legal status 167
 membership 35-7, 39, 166
 mission 27-8
 school bulletin 46
 social hierarchy 57-8
National Geographic Ventures 167
natural rights 76
The New Republic 64
Nicaragua 155
Nicaragua Canal 28
Niles, Blair 151
nude
 as ideal 78
 in *NGM* 78

Ohmann, Richard 4, 42
Orientalism, in *NGM* 21, 166
Osgood, Robert 9-10
otherness
 and identity 5, 111
 and sexuality 15

Pan American Women's Conference (1922) 150
Pan Americanism 154, 163
Pan-American Union 18, 151
ideals 155
origins 154
Panama, in *NGM* 51-2
Panama Canal 51, 155
Pankhurst, Emmeline 150
Patric, John 58
Patterson, Carolyn Bennett 138, 139
Pauly, Philip 20
Peale, Charles Wilson, 'The Artist in His Museum' 75
Peary, Robert E. 19
Peck, Annie Smith 147
Philippines 53
census 54-5, 56
nationalist movement 55-6
in *NGM* 32, 54-6, 157-8
Phillips, Richard 52
photographers
as artists 109-10
NGM instructions to 125-6
photography
and colonial control 86
and criminology 85-6
Harriet Adams 134
and imperialism 13, 56, 71, 84
justification of, Maynard Williams 106-8
in *NGM* 13, 14, 22, 54-5, 69, 89-96, 120-1, 168-9
and portraiture 83
Susan Sontag on 82, 96, 97
and typification 81-2, 84-6, 87, **91**, 89-96, 109-13
Poole, Robert M., *Explorer House* 20
portraiture, and photography 83
Pratt, Julius W. 48
Pratt, Mary Louise 7-8, 71, 73, 79-80, 100
Imperial Eyes 6
on women explorers 142
Primo de Rivera, Gen Manuel 158
Progressive Party 44
Progressivism
features 43-4
imperialist ideology 44-5
origins 44
see also liberal-developmentalism

race
Harriet Adams on 161-2
hierarchy 47-8
Reader's Digest 59
and American identity 5
Ritter, Carl 73
Roosevelt, Theodore 51, 53, 140
Corollary to Monroe Doctrine (1904) 154-5
NGM contributions 44, 46
Rose, Gillian 22
Rosenberg, Emily 48
on liberal-developmentalism 45, 46-7
Rowe, Leo Stanton 155
Royal Geographical Society (RGS) 26-7
Hints for Travellers 87
Ryan, James 13, 87

Saltonstall, Leverett 165
Samoa, in *NGA* 70
Saturday Evening Post 5
Sauer, Carl 73
Schiebinger, Londa 76
Schneirov, Matthew 4, 42
Schoedler, Lillian 116
Schulten, Susan 11, 65
The Geographical Imagination 22
Scidmore, Eliza 93, 138-9
science
'anti-conquest' narratives 7-8
and empire 8
Scribner's 42
'seeing man'
Harriet Adams as 163
Maynard Williams as 100-3, 129-30
see also anti-conquest narratives
Sekula, Allan 83, 85
Seldes, George 59-60
self-determination doctrine, Wilson 58
Sen, Gertrude Emerson 151
Seton, Ernest Thompson 140
sexuality, and otherness 15
Sharp, Joanne 5
Shelby, Gertrude 151
Simms III, John A 162
Smithson, James 28
Smithsonian Institution 28, 33

Society for the Diffusion of Useful Knowledge 27
Society of Women Geographers 18, 135
expansion 151
foundation 150-1
Sontag, Susan, on photography 82, 96, 97
South America, America, trade 155-7
Spanish-American War (1898) 11, 22
and American destiny 30-1
American economic opportunities 31
geographers' views 30
NGM opportunities 31-2
rhetoric of 29
Spencer, Herbert 47
Stafford, Barbara 73
Stassen, Howard 165
Stepan, Nancy 78
Stoddart, David 72
'strategies of innocence'
of Maynard Williams 115
in *NGM* 6, 7, 24, 169
Steet, Linda 21, 92
Sullivan, William F. 90
Sumatra 58

Taft, William H.
NGM contributions 46, 48-9
'Ten Years in the Philippines' 48-9, 53
Tarbell, Ida 139
'Teddy Bear Patriarchy' 20, 46, 52-3
Thomas, Alan 87
Thomas, Nicholas 79, 81
Tiffany, Sharon W. 14
tourism, Maynard Williams' hostility to 122-4, 130
Trachtenberg, Alan 75
travel writing
'anti-conquest' 7
as imperialism 6-7
traveling, vs exploration 143-4
Trinidad 162
Tristan, Flora 142
Tuason, Julie 20-1
Turgovnick, Marianna 94
Tylor, E.B. 84, 87
types 77-82, 89-90
ideal
construction of 79-80
Weber on 79
images **89**, **91**
typification
in geography books 81
in *NGM* 80, 82, 89-96
and photography 81-2, 84-6, 87, **91**, 89-96, 109-13

Wallace, DeWitt 59
Weber, Max, on ideal types 79
Williams, Maynard Owen 16, 17-19, 21, 61, 90, 94-5, 99-130
on African-Americans 117-18
and bare-breasted women 114-19
on beauty 109, 110
elitism 120, 121-2, 125
NGM editors, relationship 119-22
photograph of **101**
photographs by **107**, **111**, **129**
staging of 109-12, 126-7
photography, justification for 106-14
racial attitudes 117-19
as 'seeing man' 100-3, 129-30
'strategies of innocence' 115
tourism, hostility to 122-4, 130
world friendship, commitment to 104-5, 108-9
writings
'At the Tomb of Tutankhamen' 167-8
'Latvia, home of the Letts' 80
Williams, Raymond 3
Wilson, Woodrow, President, self-determination doctrine 58
Wisherd, Bud 127
World Outlook 152, 158
The World's Work 135, 158

For Product Safety Concerns and Information please contact our EU representative GPSR@taylorandfrancis.com Taylor & Francis Verlag GmbH, Kaufingerstraße 24, 80331 München, Germany